Lecture Notes Editorial Policies

Lecture Notes in Statistics provides a format for the informal and quick publication of monographs, case studies, and workshops of theoretical or applied importance. Thus, in some instances, proofs may be merely outlined and results presented which will later be published in a different form.

Publication of the Lecture Notes is intended as a service to the international statistical community, in that a commercial publisher, Springer-Verlag, can provide efficient distribution of documents that would otherwise have a restricted readership. Once published and copyrighted, they can be documented and discussed in the scientific literature.

Lecture Notes are reprinted photographically from the copy delivered in camera-ready form by the author or editor. Springer-Verlag provides technical instructions for the preparation of manuscripts. Volumes should be no less than 100 pages and preferably no more than 400 pages. A subject index is expected for authored but not edited volumes. Proposals for volumes should be sent to one of the series editors or addressed to "Statistics Editor" at Springer-Verlag in New York.

Authors of monographs receive 50 free copies of their book. Editors receive 50 free copies and are responsible for distributing them to contributors. Authors, editors, and contributors may purchase additional copies at the publisher's discount. No reprints of individual contributions will be supplied and no royalties are paid on Lecture Notes volumes. Springer-Verlag secures the copyright for each volume.

Series Editors:

Professor P. Bickel
Department of Statistics
University of California
Berkeley, California 94720
USA

Professor P. Diggle
Department of Mathematics
Lancaster University
Lancaster LA1 4YL
England

Professor S. Fienberg
Department of Statistics
Carnegie Mellon University
Pittsburgh, Pennsylvania 15213
USA

Professor K. Krickeberg
3 Rue de L'Estrapade
75005 Paris
France

Professor I. Olkin
Department of Statistics
Stanford University
Stanford, California 94305
USA

Professor N. Wermuth
Department of Psychology
Johannes Gutenberg University
Postfach 3980
D-6500 Mainz
Germany

Professor S. Zeger
Department of Biostatistics
The Johns Hopkins University
615 N. Wolfe Street
Baltimore, Maryland 21205-2103
USA

Lecture Notes in Statistics 177

Edited by P. Bickel, P. Diggle, S. Fienberg, K. Krickeberg,
I. Olkin, N. Wermuth, S. Zeger

Springer
London
Berlin
Heidelberg
New York
Hong Kong
Milan
Paris
Tokyo

Caitlin E. Buck
Andrew R. Millard (Editors)

Tools for Constructing Chronologies

Crossing Disciplinary Boundaries

With 54 Figures

Springer

Caitlin E. Buck
Department of Probability and Statistics
University of Sheffield
Hicks Building
Hounsfield Road
Sheffield S3 7RH
UK

c.e.buck@sheffield.ac.uk

Andrew R. Millard
Department of Archaeology
University of Durham
South Road
Durham DH1 3LE
UK

a.r.millard@durham.ac.uk

British Library Cataloguing in Publication Data
Tools for constructing chronologies : crossing disciplinary
    boundaries. - (Lecture notes in statistics)
    1. Chronology – Statistical methods 2. Archaeology –
    Methodology 3. Paleoecology – Methodology 4. Bayesian
    statistical decision theory
    I. Buck, Caitlin E. II. Millard, Andrew
    930.1'015195
    ISBN 185233763X

Library of Congress Cataloging-in-Publication Data
Tools for constructing chronologies : crossing disciplinary boundaries / [edited by] Caitlin
    E. Buck, Andrew R. Millard.
        p.   cm. -- (Lecture notes in statistics ; 177)
    Includes bibliographical references and index.
    ISBN 1-85233-763-X (acid-free paper)
    1. Chronology - Statistical methods. 2. Bayesian statistical decision theory. I. Buck,
Caitlin E. II. Millard, Andrew. III. Lecture notes in statistics (Springer-Verlag) ; v. 177.
CE12.T66   2004
902'.02—dc22   2003061219

ISBN 1-85233-763-X Springer-Verlag London Berlin Heidelberg
Springer-Verlag is a part of Springer Science+Business Media
springeronline.com

Typesetting: Camera-ready by the editors
12/3830-543210 Printed on acid-free paper SPIN 10937945

Preface: Towards Integrated Thinking in Chronology Building

Robust and reliable chronologies are essential to many research projects in the historical sciences (e.g. archaeology, palaeoecology, geology), history, and other disciplines with historical aspects (e.g. evolutionary genetics), as well as in heritage management and petrochemical exploration. The nature of the methods in use in these disciplines varies quite considerably and in the past there has been very little cross-fertilisation of ideas between them. Despite this, all the methods have one thing in common: they take a collection of *dates* or *temporal relationships* for a series of individual events and combine them with other information to synthesise a *chronology* which may include the inferred dates of events for which no direct dating evidence is available. The methods used range from formal mathematical and statistical models (e.g. Agterberg and Gradstein 1988; Buck et. al. 1996) to less formal and more subjectively argued approaches (e.g. Gasche et al. 1998; Holst 1999). In some cases very similar methods have been developed independently. For example, similar mathematical models for estimating the unknown age of a boundary from stratigraphically related radiometric dates have been developed for archaeology (Buck et al. 1992; Zeidler et al. 1998), biostratigraphy (Agterberg and Gradstein 1988), Quaternary science (Biasi and Weldon 1994) and the geomagnetic polarity timescale (Agrinier et al. 1999), but in each case using a different computational method. Examples of cross-disciplinary transfer of methods are quite rare, but do seem to be on the increase. One example would be the work of Blackham (1998) who encourages archaeologists to adopt relative dating techniques (known as unitary association) that are already routinely used in biostratigraphy. Another would be recent collaborative work by Buck et al. (2003) to demonstrate the utility of Bayesian radiocarbon calibration methods in tephrochronology as well as archaeology.

The dates and other information used to build chronologies are usually costly and time consuming to collect. Often they require many hours of painstaking work both in the field and laboratory, and they utilise expensive and specialised equipment. The creation of a chronology often requires that thousands of pounds (or dollars or euros) are spent on obtaining data such as radiocarbon dates, stratigraphic sequences, identified and quantified collections of fossils or artefacts, ancient DNA sequences, tree-ring dates, volcanic tephra compositions and/or architectural phase information. Each year government and private funding bodies around the world invest very large sums both in hardware for chronometric dating and in individual research projects that generate enormous amounts of data which need interpreting.

Typically, when researchers have problems making all the chronological information fit together, they focus on trying to improve the quality of the absolute and relative dating evidence available to them. While this is clearly beneficial it is also expensive. For example, halving the error limits on a radiometric age requires at least four times the number of determinations, or four times as much counting time, and the exercise of "chronometric hygiene" (Spriggs 1989) can result in the rejection of numerous dates as useless. It is readily shown that in many cases there are alternative and more effective ways to achieve a satisfactory synthesis (for example Christen 1994; Buck and Christen 1998; Philip and Millard 2000).

This book takes a different approach: it focuses on ways to get more out of the existing chronological data by careful analysis. Although the editors' research in chronology building focuses on adoption of the Bayesian statistical paradigm (and we are motivated by the fact that we believe that this could fruitfully have much wider application), there are also many other methods that are presently confined to one discipline and might be more widely applied. During the last decade new chronology building techniques have been developed in a number of disciplines and we considered that the time was right for researchers from these different backgrounds to share experiences, pool resources and reflect on the future of research in formal chronology building. As a result, we obtained joint funding from the Natural Environment Research Council (UK), English Heritage and the Wenner-Gren Foundation (USA) to hold a workshop and put together this book. In compiling this volume we have sought authors from a wide range of disciplines who are conducting cutting edge research in formal chronological methods, and via the workshop and editing of the material have tried to provide an integrated and cross-linked volume.

The specially invited chapters survey a range of leading edge methods in chronology construction and seek to enable cross-disciplinary fertilisation of techniques. They cover a range of timescales (from trilobites to HIV) from the perspectives of a number of disciplines (including geology, archaeology, and statistics). The methods used range from treatments utilising complex statistical models (e.g. of archaeomagnetic calibration and gene-trees) to considerations of how systematically to represent relative dating information (e.g. in trilobite extinctions and Iron Age villages).

Each chapter can be read on its own, but cross-references to parallels and contrasts in other chapters are included. This preface and the editors' introductory summary of each chapter provide an overview of the state of chronology construction methods across a range of disciplines and highlight links between them. The first chapters focus on the state of the art in methodology (Buck, Chapter 1) and applications (Bayliss and Bronk Ramsey, Chapter 2) of Bayesian approaches to integrating radiocarbon dates and stratigraphic information. Approaches to the integration of large amounts of chronological information are considered with respect to the analysis of stratigraphy from numerous features on single archaeological sites (Holst, Chapter 6), and with respect to the archaeology of the entire Eastern Mediterranean in the second millennium BC (Cichocki et al., Chapter 4). The latter includes tephrochronological data, which is also treated at more length with respect to methodology and application in the North Atlantic region (Dugmore et al., Chapter 8). The integration of different sources of chronological information is also considered with respect to dating sea-level changes (Edwards, Chapter 9) and the construction of an archaeomagnetic reference curve (Lanos, Chapter 3). The Bayesian statistical analysis of chronological data is the major theme of the remaining chapters of the book, covering chronometric techniques other than radiocarbon (Millard, Chapter 11), choice of relative dating models in statistical chronology building (Sahu, Chapter 5), a statistical framework for analysing fossil sequences (Weiss et al., Chapter 10), and the reconstruction of genealogies from genetic sequences of differing dates (Drummond et al., Chapter 7).

Our intention is for this volume to reach an interdisciplinary audience. Whilst editing it we have become very aware of the extent to which jargon and technical language are a barrier to the transfer of ideas between disciplines. We have attempted to ensure that specific terms occurring in only a few places are explained at their first occurrence, but two groups of terms and concepts recur throughout the book, so we provide definitions here. The first group are statistical, reflecting the fact that many of the authors in the volume have chosen to adopt modern statistical tools to help in the chronology building process. The second group are used by chronologists but may be unfamiliar to our statistical readers. Here we give brief, informal explanations to aid understanding. Readers requiring formal definitions should consult specialist dictionaries of statistics and archaeology.

Statistical terms and approaches

In statistics there are two broad methodological schools: Classical and Bayesian. **Classical** statistics makes inferences by considering the likelihood of obtaining a particular set of observations given the values of the parameters in a probability model. In contrast a **Bayesian approach** makes inferences based on the *a posteriori* probability distributions of the parameters as given by Bayes' theorem, which combines *a priori* probabilities for the parameters with the likelihood of the data. In a Bayesian approach all forms of uncertainty are expressed in terms of probability, and what we know before we collect new data (i.e. *a priori* information) is held to be essential to understanding.

More formally, Bayes' theorem is given by

$$p(parameters|data) = \frac{p(data|parameters) \times p(parameters)}{p(data)}$$

where p(parameters) expresses our prior beliefs about the values of parameters before obtaining data, p(data|parameters) is the likelihood of obtaining our observations if the parameters were known, p(data) is the probability of obtaining the data summed (or integrated) over all possible values of the parameters and p(parameters|data) expresses our posterior beliefs incorporating our prior beliefs and the data.

The distribution of probability over possible values of parameters or data is formally expressed as a **probability density function** (often called a **density function** or even just a **density**). This is a mathematical function such that, if it is plotted on a graph, the area between the curve and the x-axis is one (with the area under the curve representing probability). This is an important feature, because probabilities are recorded on the scale zero to one and when all possibilities are taken into account their probabilities must always add up to one.

The **likelihood function** (often simply called a **likelihood**) is the form taken by a particular probability model (often a probability density function) given that specific data relating to the parameters of that model have been observed. Put simply, a likelihood describes the probability of obtaining a particular set of data, given the probability models we have selected to represent the phenomena under study and specific values of their parameters.

A priori is a Latin term used to identify information or knowledge which is held before observations are made and is used interchangeably with the term **prior information**. Such information is usually formulated as a single numerical statement or as a probability density function (see above) and must be elicited from an appropriate expert or via study of relevant literature. The terms uninformative, non-informative or vague prior are used in situations where there is little or no prior information available.

Similarly, *a posteriori* is used to identify information or knowledge which is held after observations are made. Typically, in Bayesian statistics, this term is used interchangeably with the term **posterior information**. Bayesian statisticians derive posterior information by combining prior information, a likelihood function and relevant data.

Elicitation is the process of obtaining and summarising expert knowledge about some unknown quantity of interest, which can then be used to provide a prior probability in a Bayesian analysis. If the expert in question does not have a statistical background, as is often the case, translating their beliefs into a statistical form suitable for use in Bayesian analyses can require careful thought. Hence there is also often a need for sensitivity checking to be sure that we understand what effect a particular probabilistic representation of prior information is having on the posterior inferences obtained.

The computation and results of statistical methods also have their own jargon. Complex statistical models cannot be solved algebraically, so a method based on numerical simulations called **Markov chain Monte Carlo** simulation (or **MCMC**) is used. MCMC provides enormous scope for realistic statistical modelling. Until recently, without such methods, allowing for the full complexity and structure in Bayesian models was practically impossible. Rather than attempting to do the calculations exactly, we simulate from the required distributions and then use the samples we obtain to draw conclusions about the posterior distributions of interest. Provided that our samples are sufficiently large, then, subject to certain conditions, they provide robust and reliable estimates of the posterior distributions of interest. The most important condition is that the sequence of simulated values has reached a state known technically as **convergence**, where the summarised results are independent of the starting point of the sequence.

The results of a Bayesian analysis are often summarised as a **highest posterior density region** (or **HPD** region), which is the shortest interval that can be constructed to contain a particular fixed percentage (typically 95%) of a posterior density. As well as containing a fixed percentage of the probability, HPD intervals are constructed in such a way that all those values inside the range are more likely than those outside it. For multimodal distributions, such an interval need not be continuous (in other words, an HPD region may be composed of several separate sub-intervals). This is particularly common in radiocarbon dating, where the 'wiggles' in the calibration curve can induce deep troughs in posterior probability densities. Such troughs have a very low probability of representing the true calendar date while dates on either side are much more likely.

Chronological terms and approaches

Chronological data and the chronologies constructed from them come in a variety of forms. One common way of dividing them is according to the placement of events in time: relative chronologies specify only the order of events and nothing about their position in absolute (i.e. numerical) time; floating chronologies specify the time lapses between events but do not specify when in absolute time they are; and absolute chronologies specify the position of events in numerical time with respect to the present. We note that, elsewhere, some authors equate high precision or exact dates with "absolute dates", but this is not the usage in this volume. Here we distinguish the *type* of chronological information from the *uncertainty* of that information.

Two frequently used chronological terms may need explanation for non-specialists. The Latin term *terminus ante quem* (or TAQ) can be translated "date before which". The complementary term *terminus post quem* (or TPQ) translates as "date after which". They are used to indicate the date before (or after) which a particular past event must have taken place. TAQs and TPQs are typically established by the identification of stratigraphic relationships between objects or layers found during excavation or coring, but can also be established on the basis of historical evidence of major events such as earthquakes or fires.

The current volume

We are aware that this volume is not comprehensive in its coverage of chronological issues. The range of tools in archaeology alone is much richer than that reported here. In particular we do not have full coverage of typology, and its application in seriation (e.g. Wilkinson 1996; O'Brien and Lyman 1999). Seriation is of course one of the oldest formal methods in archaeology, dating back to the start of the 20th century and the paper-based methods of Petrie (1899) in ordering graves by their contents to form "battleship curves" showing the rise and decline of different shapes and styles of artifacts. However, as illustrated by Laxton (1990), Halekoh and Vach (1999), Buck and Sahu (2000) and as hinted at by both Buck and Sahu in their chapters for this volume, there is still more work to do.

Another major omission is historical chronologies. Often these are considered to be unproblematic and precise, which may be true in well-documented periods, but as we reach further into the past and approach the border between archaeology and history, we find that there is considerable uncertainty. This is one of the major drivers behind the SCIEM2000 project (Cichocki et al., Chapter 4). It is also important in the elucidation of ancient kinglists and related archaeological information (see, for example, Gasche et al. 1998) which has its own methods that are not discussed here, but could fruitfully be linked to some of the approaches described.

In addition to the techniques described by Dugmore et al. (Chapter 8) and Weiss et al. (Chapter 10), geology also has a number of formal and quantitative methods not covered here, including quantitative stratigraphy (Agterberg and Gradstein 1988) for stratigraphic correlation, sequence slotting and other procedures for correlating sequences of species assemblages (Kovach 1993), correlation of oxygen isotope stratigraphies, laminated sediments and computed orbital parameters of the Earth (Hilgen et al. 2000; Hinnov 2000). Once archaeologists and geologists, in particular, start to talk more widely about the tools they use for constructing chronologies, there are undoubtedly further links to be made between those already in use in geology and those described here and therefore further potential for interdisciplinary transfer of ideas.

The contents of this volume reflect the editors' knowledge of current work, and contacts, within the field of chronology building, which is why there may seem to be a bias towards Bayesian techniques and archaeology. Most of the Bayesian chapters and some of the non-Bayesian chapters share the common problem of combining qualitative information (e.g. stratigraphic ordering) with quantitative information (e.g. chronometric determinations or genetic sequences). Within a Bayesian context this is done by specifying the qualitative information as a prior constraint on the dates of events, in the other chapters it is treated less formally, either because their is no need to quantify it (e.g. in determining which of several tephras a particular deposit contains) or because qualitative inferences are the aim of the analysis (e.g. in determining the trend in sea-level). Holst's chapter differs from the others in using only relative chronological data, and, in contrast to the practice of some of the Bayesian analyses, allowing for its inherent uncertainty. Clearly, there is often uncertainty in relative data (as Bayliss and Bronk Ramsey show in Chapter 2) and a focus for future research could be how, formally and quantitatively, to combine relative dating and absolute dating, whilst accounting for the uncertainty in both aspects. In a sense this is what Drummond et al. (Chapter 7) do when they account for uncertainty in the topology of the gene-tree, as well as in the dating of the events in the gene-tree.

It would appear that quantitative statistical methods have seen greater development in archaeology, perhaps because the questions archaeologists ask are finer-grained than those of geologists or geneticists. However, the chapters here by Edwards, Millard, Weiss et al. and Dugmore et al. all show that high temporal resolution is increasingly sought in geological studies. This is nowhere more true than in palaeoclimate studies, where the ice-core records give annual resolution and show massive glacial to inter-glacial temperature changes happening over decadal or sub-decadal timescales, whilst traditional chronologies have been given with a resolution expressed in centuries. It is now recognised that sophisticated statistical approaches will be necessary to date and correlate terrestrial palaeoecological sequences with the ice-cores (Sarnthein et al. 2002) in order to understand the causes and impacts of past climate-change.

We hope that this volume will stimulate researchers from various disciplines to think about the adaptation and application of approaches from other disciplines. The participants in the workshop were pleasantly surprised by the extent of overlap between their interests and found that crossing disciplinary boundaries was nowhere near as difficult as some had feared before they met. All the chapters in this volume offer us insights into the approaches we need to take to achieve integrated thinking in chronology building and illustrate both the nature of the methods that are needed to achieve it and the rewards that can be gained. Several of the contributors have also found new collaborators (or reinforced existing ones) as a result of the workshop and it seems likely that this area of research will continue to be a lively and interesting one.

Acknowledgements

We would like to thank those people and organisations who have helped us in this project. Without generous funding from NERC (Connect A Grant reference NER/E/S/2001/00396), the Wenner-Gren Foundation (workshop grant reference G. CONF-346) and English Heritage it would never have happened.

Alex Bayliss encouraged us to set up the conference and enabled us to obtain English Heritage support. The staff of the University of Wales Conference Centre at Gregynog provided an excellent environment away from all our normal worries where we could concentrate on chronologies. The authors of the chapters have of course produced what you see before you here, but the quality of their work has been enhanced by the discussants at the workshop, Marion Scott, Tony O'Hagan, Steve Roskams and Tom Higham (who provided written comments to the authors), by the notes of Delil Gomez-Portugal-Aguilar taken at the time of the meeting and by the comments from a number of anonymous referees to whom we are all extremely indebted. We are also indebted to the editorial team at Springer-Verlag (in particular Stephanie Harding and Karen Borthwick), to David Cruikshank at the Applied Probability Trust who helped us with LATEX queries and to Kathleen Lyle who compiled the index for the volume. Finally, as editors and organisers we have been supported by our partners Geoff and Julia: we thank them especially for their forbearance as we deserted them at times in order to construct chronologies.

Sheffield and Durham,   *Caitlin Buck*
May 2003   *Andrew Millard*

References

Agrinier, P., Gallet, Y. and Lewin, E. (1999). On the age calibration of the geomagnetic polarity timescale. *Geophysical Journal International*, **137**, 81–90.

Agterberg, A. P. and Gradstein, F. M. (1988). Recent developments in quantitative stratigraphy. *Earth Science Reviews*, **25**, 1–73.

Biasi, G. P. and Weldon, R. (1994). Quantitative refinement of calibrated C-14 distributions. *Quaternary Research*, **41**, 11–18.

Blackham, M. (1998). The unitary association method of relative dating and its application to archaeological data. *Journal of Archaeological Method and Theory*, **5**, 165–207.

Buck, C. E., Cavanagh, W. G. and Litton, C. D. (1996). *Bayesian approach to interpreting archaeological data*. John Wiley, Chichester.

Buck, C. E. and Christen, J. A. (1998). A novel approach to selecting samples for radiocarbon dating. *Journal of Archaeological Science*, **25**, 303–310.

Buck, C. E., Higham, T. F. G. and Lowe, D. J. (2003). Bayesian tools for tephrochronology. *Holocene*, **13**, in press.

Buck, C. E., Litton, C. D. and Smith, A. F. M. (1992). Calibration of radiocarbon results pertaining to related archaeological events. *Journal of Archaeological Science*, **19**, 497–512.

Buck, C. E. and Sahu, S. K. (2000). Bayesian models for relative archaeological chronology building. *Applied Statistics*, **49**, 423–440.

Christen, J. A. (1994). Summarizing a set of radiocarbon determinations: a robust approach. *Applied Statistics*, **43**, 489–503.

Gasche, H., Armstrong, J. A., Cole, S. W. and Gurzadyan, V. G. (1998). *Dating the fall of Babylon: a reappraisal of second-millennium chronology*. Mesopotamian history and environment, Series II, Memoirs, vol. 4. University of Ghent and the Oriental Institute of the University of Chicago, Chicago, IL.

Halekoh, U. and Vach, W. (1999). Bayesian seriation as a tool in archaeology. In *Archaeology in the age of the Internet: proceedings of the conference on computer applications and quantitative methods in archaeology, 1997*. Archaeopress: British Archaeological Reports, Oxford, International Series, **S750**, 107. Complete paper on CD with volume.

Hilgen, F. J., Bissoli, L., Iaccarino, S., Krijgsman, W., Meijer, R., Negri, A. and Villa, G. (2000). Integrated stratigraphy and astrochronology of the messinian GSSP at Oued Akrech (Atlantic Morocco). *Earth and Planetary Science Letters*, **182**, 237–251.

Hinnov, L. A. (2000). New perspectives on orbitally forced stratigraphy. *Annual Review of Earth and Planetary Sciences*, **28**, 419–475.

Holst, M. K. (1999). The dynamic of the Iron-age village: a technique for the relative-chronological analysis of area-excavated Iron-age settlements. *Journal of Danish Archaeology*, **13**, 95–119.

Kovach, W. L. (1993). Multivariate techniques for biostratigraphical correlation. *Journal of the Geological Society*, **150**, 697–705.

Laxton, R. (1990). Methods of chronological ordering. In A. Voorrips and B. Ottoway (eds.), *New Tools from Mathematical Archaeology*, Scientific Information Centre of the Polish Academy of Sciences, Warsaw.

O'Brien, M. J. and Lyman, R. L. (1999). *Seriation, stratigraphy and index fossils: the backbone of archaeological dating*. Kluwer Academic / Plenum Publishers, New York.

Petrie, W. M. F. (1899). Sequences in prehistoric remains. *Royal Anthropological Institute of Great Britain and Ireland Journal*, **29**, 295–301.

Philip, G. and Millard, A. R. (2000). Khirbet Kerak Ware in the Levant: the implications of radiocarbon chronology and spatial distribution. In C. Marro and H. Hauptman (eds.), *Chronologies des pays du Caucase et de l'Éuphrate aux IVème-IIIème millénaires*, Boccard, Paris, Institut Français d'études anatoliennes d'İstanbul, Varia Anatolica VIII, 279–296.

Sarnthein, M., Kennett, J. P., Allen, J. R. M., Beer, J., Grootes, P., Laj, C., McManus, J., Ramesh, R. and SCORIMAGES Working Group (2002). Decadal-to-millennial-scale climate variability – chronology and mechanisms: summary and recommendations. *Quaternary Science Reviews*, **21**, 1121–1128.

Spriggs, M. (1989). The dating of the Island Southeast Asian Neolithic: an attempt at chronometric hygiene and linguistic correlation. *Antiquity*, **63**, 587–613.

Wilkinson, T. A. H. (1996). *State formation in Egypt: chronology and society*. Cambridge Monographs in African Archaeology, **40**, British Archeological Reports International Series, **S651**. Tempus Reparatum, Oxford.

Zeidler, J. A., Buck, C. E. and Litton, C. D. (1998). The integration of archaeological phase information and radiocarbon results from the Jama River Valley, Ecuador: a Bayesian approach. *Latin American Antiquity*, **9**, 135–159.

Contents

XVI   Contents

List of Contributors

Sanjib Basu
Department of Statistics, Northern Illinois University, De Kalb, IL 60115, USA. basu@math.niu.edu

Alex Bayliss
English Heritage, Centre for Archaeology, 23 Savile Row, London W1S 2ET, UK. alex.bayliss@english-heritage.org.uk

Max Bichler
Institut für Astronomie, Universität Wien, Türkenschanzstraße 17, A-1180 Vienna, Austria.

Caitlin E. Buck
Department of Probability and Statistics, University of Sheffield, Hicks Building, Hounsfield Road, Sheffield S3 7RH, UK. c.e.buck@sheffield.ac.uk

Christopher Bronk Ramsey
Oxford Radiocarbon Accelerator Unit, Research Laboratory for Archaeology and the History of Art, Oxford University, 6 Keble Road, Oxford OX1 3QJ, UK. christopher.ramsey@rlaha.ox.ac.uk

Otto Cichocki
Dendrolab – VIAS – Universität Wien, Althanstraße 14, A-1090 Vienna, Austria. otto.cichocki@univie.ac.at

Gertrude Firneis
Institut für Astronomie, Universität Wien, Türkenschanzstraße 17, A-1180 Vienna, Austria.

Alexei Drummond
School of Biological Science, Auckland University, Private Bag 92019, Auckland, New Zealand.

Andrew J. Dugmore
Department of Geography, University of Edinburgh, Edinburgh EH8 9XP,
UK. ajd@geo.ed.ac.uk

Robin J. Edwards
Departments of Geography and Geology, Trinity College Dublin, Dublin 2,
Ireland. edwardsr@tcd.ie

Mads Kähler Holst
Department of Prehistoric Archaeology, University of Aarhus, Moesgaard,
8270 Højbjerg, Denmark. farkmh@hum.au.dk

Walter Kutschera
Institut für Radiumforschung und Kernphysik, Universität Wien, VERA
Laboratorium, Währinger Straße 17, A-1090 Vienna, Austria.

Philippe Lanos
Laboratoire d'Archéomagnétisme, UMR 6566 du CNRS Civilisations
Atlantiques et Archéosciences, Université de Rennes 1, Géosciences-
Rennes, Campus de Beaulieu, bât. 15, 35042 Rennes, France.
philippe.lanos@univ-rennes1.fr

Guðrún Larsen
Department of Geography, University of Edinburgh, Edinburgh EH8 9XP,
UK and Science Institute, University of Iceland, Reykjavik, IS-101, Iceland.

Charles R. Marshall
Museum of Comparative Zoology, Harvard University, 26 Oxford Street,
Cambridge, MA 02138, USA. cmarshall@oeb.harvard.edu

Andrew R. Millard
Department of Archaeology, University of Durham, South Road, Durham
DH1 3LE, UK. a.r.millard@durham.ac.uk

Wolfgang Müller
Österreichisches Archäologisches Institut, Zweigstelle Kairo, Zamalek, Sharia
Ismail Muhammed Apt. 62/72, ET-Kairo, Egypt.

Anthony J. Newton
Department of Geography, University of Edinburgh, Edinburgh, EH8 9XP,
Scotland, UK.

Geoff K. Nicholls
Department of Mathematics, Auckland University, Private Bag 92019,
Auckland, New Zealand. nicholls@math.auckland.ac.nz

Allen G. Rodrigo
School of Biological Science, Auckland University, Private Bag 92019, Auckland, New Zealand.

Sujit K. Sahu
Faculty of Mathematical Studies, University of Southampton, Highfield, Southampton SO17 1BJ, UK. s.k.sahu@maths.soton.ac.uk

Wiremu Solomon
Department of Mathematics, Auckland University, Private Bag 92019, Auckland, New Zealand.

Peter Stadler
Naturhistorisches Museum Wien, Praehistorische Abteilung, Burgring 7, A-1010 Vienna, Austria.

Robert E. Weiss
Department of Biostatistics, UCLA School of Public Health, Los Angeles, CA 90095-1772, USA. robweiss@ucla.edu

1

Bayesian Chronological Data Interpretation: Where Now?

Caitlin E. Buck

Summary. This chapter summarizes the current state of the art in the use of Bayesian statistical models originally devised for chronology building in archaeology. Over the last 10 years, archaeologists have begun routinely to adopt such methods based on these models because they offer a formal, coherent framework for the integration of chronometric (typically radiocarbon) data and other sources of absolute and relative chronological information. Since they allow us to combine both relative and absolute dating evidence from different sources, such methods typically lead to less uncertainty in the final date estimates than other methods which focus on one piece of evidence at a time. Until recently, such methods have only routinely been used by archaeologists, but many of them clearly have potential for chronology building in other disciplines too. Counterpoint to this chapter is given by the discussions of the practical, routine application of its methods by Bayliss and Bronk Ramsey (Chapter 2). In addition, further developments to some of the ideas here are discussed by Lanos (Chapter 3) with respect to archaeomagnetism, Sahu (Chapter 5) with respect to model choice and Millard (Chapter 11) with respect to techniques other than radiocarbon.

1.1 Introduction

Bayesian statistical tools are now routinely used by practitioners in many applied sciences who wish to formally, and coherently, incorporate *a priori* knowledge, a formal model and suitable data to arrive at (*a posteriori*) inferences based on all three. Archaeology was one of the early subjects to gain tools in this area when, more than a dozen years ago, Bayesian statisticians first started to take an interest in helping develop tools for formal chronology building (Naylor and Smith 1988). Initially, attention focused on the calibration and interpretation of radiocarbon data. Very quickly, however, it became clear that the framework was ideally suited to the integration of chronological information from a range of different sources – in particular stratigraphic sequences and historical or absolute dating evidence (Buck et al. 1991, 1992,

2   Caitlin E. Buck

1994a,b,c; Buck and Litton 1995; Litton and Buck 1996; Buck and Christen 1998).

In a separate strand of research, a Bayesian approach to relative chronological data interpretation (in the absence of chronometric evidence) has also been suggested (Buck and Litton 1991). This work showed that it is possible (although computationally intensive) to devise Bayesian versions of traditional seriation models, thus allowing *a priori* relative chronological information to be incorporated into the data interpretation process. In more recent work in this area (Buck and Sahu 2000), we have also been able to identify model choice tools which mean that we can consider a range of options for modelling a) the underlying processes that gave rise to the archaeological deposits we excavate today, and b) the *a priori* information that is so often vital to coherent interpretation.

Two Bayesian chronology building packages are already widely used by archaeologists: *OxCal* (Bronk Ramsey 1995) and *BCal* (Buck et al. 1999). Both assume that the user's primary interest is in the interpretation of radiocarbon data. They allow users with relatively little statistical knowledge to build quite sophisticated chronologies, tying together chronological information in the form of:

- chronometric data,
- *a priori* information about calendar dates,
- stratigraphic sequences,
- archaeological phases, and
- interrelated phases and sequences.

The framework thus implemented is general, scalable and eminently well suited to the addition of further tools to aid in chronology building. In recent years a number of researchers have noted this and have extended the framework to help to solve particular problems. In this chapter I am going to summarize the Bayesian chronological data interpretation framework as currently used, review recent and on-going work that offers extensions which are not yet routinely in use, and discuss what might usefully be done in the future. I will not, however, provide an introduction to the Bayesian philosophy of science or discuss statistical implementation in any detail. Both of these topics are covered in Buck et al. (1996) and in other references cited in this chapter.

1.2 What We Can Already Do

In this section I look at the kinds of chronology building problems that are already routinely tackled within the Bayesian framework. Such problems typically have radiocarbon data at their core and the framework I describe was originally motivated by the fact that such data need to be calibrated in order to obtain estimates of the calendar date of events of interest.

1.2.1 Calibrate Single Radiocarbon Determinations

Let θ represent the unknown calendar date of a past event in calendar years before present (cal. BP)[1]. Suppose that this event can reliably be associated with a suitable organic sample for radiocarbon dating. In addition to a true underlying calendar date, this sample also has a true underlying radiocarbon age which is not unique to that sample, but is associated with all samples whose true underlying calendar date is θ. To indicate this close relation between the two, we use $\mu(\theta)$ to represent the true underlying radiocarbon age before present (BP) of a sample whose true calendar age is θ cal. BP. $\mu(\theta)$ relates to the proportion of radioactive carbon (^{14}C) present in the sample when it is analysed by the radiocarbon dating laboratory. For many reasons (not least because of the random nature of radioactive decay, and the small proportion of ^{14}C relative to the stable isotopes ^{12}C and ^{13}C), the radiocarbon dating laboratories do not give us $\mu(\theta)$ precisely. They provide an estimate of $\mu(\theta)$ which we represent using x. If the laboratory were to repeat the analysis we would be likely to get a different value for x. Thus, it is useful to see x as being just one realization of a random variable X, where $X = \mu(\theta) + \text{noise}$.

Conventionally, the noise is assumed to have a normal distribution with mean zero and standard deviation σ so that X is modelled using a normal distribution with mean $\mu(\theta)$ and variance σ^2 which is formalized thus

$$X \sim N(\mu(\theta), \sigma^2). \tag{1.1}$$

In addition to the estimate of x, radiocarbon dating laboratories also provide us with an estimate of σ so that the only other thing that we need in order to use Equation 1.1 is $\mu(\theta)$. Since the proportion of radioactive carbon in the Earth's atmosphere has not remained constant over time, and the factors which affect levels of production are not well understood, we do not know $\mu(\theta)$ precisely and cannot reliably model it. What we do have are internationally agreed calibration data (Stuiver et al. 1998) which provide pairs of estimates of X and θ from suitable samples of known age (such as tree-ring dated wood samples). These calibration data can be used to plot a calibration curve from which we can estimate θ for radiocarbon determinations from samples of unknown age. More formally, we can learn about the calendar date of a particular archaeological event by obtaining a suitable radiocarbon determination and using a statistical function (known as a *likelihood*) which relates θ to x, σ and $\mu(\theta)$. Since Equation 1.1 comprises a normal distribution, we formalize the appropriate likelihood thus

$$l(\theta; x, \sigma) \propto \exp\left\{-\frac{(x - \mu(\theta))^2}{2\sigma^2}\right\}. \tag{1.2}$$

---

[1] Note that there is a convention in radiocarbon dating that all ages are quoted before present, where 0 cal. BP is taken to be 1950 AD. Thus, on the BP timescale, all older dates are larger than more recent ones.

In practice, different researchers model $\mu(\theta)$ in different ways and hence get different graphs of the relationship between X and θ (i.e. different calibration curves). We will say more about methods for modelling the calibration data below, but for the moment, think of $\mu(\theta)$ as being constructed by a series of straight lines joining the means of the estimates of X and θ in the internationally agreed calibration data set (Stuiver et al. 1998). A calibration curve that is created in this way is known as a piece-wise linear curve. This is the form in which the calibration data have most routinely been used and results in a calibration curve that is non-linear and non-monotonic – often referred to as 'wiggly'. The 'wiggly' nature of the calibration curve results in asymmetric and, typically, multimodal calendar date estimates when Equation 1.2 is used to estimate the calendar date of a sample on the basis of new radiocarbon evidence.

Now, if each radiocarbon determination is seen as standing in isolation from others, this is all that is needed to obtain an estimate of the calendar date (θ) for the sample in question. We simply assume that a priori the value of θ is equally likely to lie anywhere in the range of the calibration data. Then our estimate for θ is based only on the radiocarbon determination (x and σ) obtained from the laboratory, the calibration curve and the form of the likelihood (in Equation 1.2). However, for many research projects, we have more than one radiocarbon determination relating to a particular period of study. Very often, we have information from stratigraphic sequences that allows us to relate the calendar dates of the radiocarbon samples one to another before any chronometric dating evidence has been obtained. Ideally, we would wish this a priori information, concerning established relative chronologies, to have a bearing on the absolute dates obtained from radiocarbon calibration. We can do this in one of two ways;

- we can create a formal framework for including information from a range of sources, or
- we can attempt to draw the various strands together in a rather more heuristic fashion – hoping that we are able to place appropriate weight on each piece of evidence we have.

Some researchers do still use the latter approach, but the former is also available via extensions to our Bayesian models as follows.

1.2.2 Calibrate Sequences of Radiocarbon Determinations

Suppose that we have several radiocarbon determinations from related past events so that
$$\theta = \{\theta_1, \theta_2, \ldots, \theta_N\},$$
$$x = \{x_1, x_2, \ldots, x_N\},$$
and
$$\sigma = \{\sigma_1, \sigma_2, \ldots, \sigma_N\}.$$

Suppose that our *a priori* information is of a relatively simple and common type, namely that we know from stratigraphic information that the calendar dates of certain events are ordered. In other words, we believe before any radiocarbon determinations are purchased that $\theta_1 < \theta_2 < \theta_3$ (cal. BP).

Given the form of the calibration curve, this kind of *a priori* information typically leads to models that cannot be solved algebraically. As a result, most software that implements such models makes use of simulation-based statistical inference tools known as Markov chain Monte Carlo (MCMC) methods. MCMC implementation details are given in Gilks et al. (1996) in which Litton and Buck (1996) provide algorithms that are particularly suited to radiocarbon calibration problems. Such methods allow us to tie the likelihood and *a priori* information together relatively easily and, for example, to restrict the estimates for θ to conform to the prior ($\theta_1 < \theta_2 < \theta_3$).

1.2.3 Include *A Priori* Information about Deposition Rates

An additional type of *a priori* information, which sometimes arises when we are interpreting sequences of radiocarbon determinations, consists of knowledge about the rate at which the samples within the sequence were deposited. In order to be legitimate *a priori* information, this knowledge must not be derived from the radiocarbon data themselves. Typical suitable prior information falls into two groups: that which provides information about the length of time elapsed between deposits, and that which allows one to posit a model for the rate of deposition of new material.

A Priori Information about Time Elapsed Between Deposits

For very many reasons, associated with the ways in which humans manipulate their environment (intentionally or unintentionally), archaeologists do not usually postulate formal models for the accumulation rate of the deposits they excavate. They do, however, sometimes have information about the likely time elapsed between the deposition of each sample in a sequence of radiocarbon dated material. This is particularly common when several radiocarbon measurements have been made on a piece of wood (or some other material with annual growth layers) so that the time elapsed between the rings or layers can be estimated (or sometimes fixed exactly). In these situations, we seek to match the whole sequence of radiocarbon age estimates from our samples to the radiocarbon calibration curve. Since the curve is often described as 'wiggly', this approach has come to be known as 'wiggle matching'.

Christen (1994a) and Christen and Litton (1995) suggested and implemented a Bayesian approach to wiggle matching that can be summarized as follows. Suppose that we have a sequence of n samples, and let the calendar dates of these samples be represented by $\theta = (\theta_1, \theta_2, \ldots, \theta_n)$. Assume also that the calendar dates are ordered so that $\theta_1 < \theta_2 < \ldots < \theta_n$ cal. BP. x_i then represents the estimated radiocarbon age of the ith sample and σ_i the corresponding standard deviation. Now, since we also have information about the time interval elapsed between each θ_{i-1} and θ_i, we represent $\theta_i - \theta_{i-1}$ by γ_i where $i = 2, 3, \ldots, n$ (γ_1 is defined as zero). Thus, in order to learn about all the calendar dates of interest, we simply need to make inferences about θ_1 and $\gamma = (\gamma_2, \gamma_3, \ldots, \gamma_n)$, since the values of the other members of θ can be obtained from them.

Given this last observation, we can represent θ_i in terms of θ_1 and γ thus

$$\theta_i = \theta_1 + \sum_{j=1}^{i} \gamma_i$$

for $i = 1, 2, \ldots, n$. As a consequence, building on the likelihood given in Equation 1.2 (and assuming that we continue to use the calibration data in the same way),

$$l(\theta_1, \gamma; x) \propto \prod_{i=1}^{n} \exp\left\{ -\frac{(x_i - \mu(\theta_i))^2}{2\sigma_i^2} \right\}. \tag{1.3}$$

This is not quite the form given in Christen (1994a) and Christen and Litton (1995) since they suggest a simultaneous remodelling of the calibration curve to take account of the errors in the curve data. We will consider such remodelling separately below.

Typically, in practical uses of models of this type, we have relatively little prior information about θ_1, but our prior information about each of the γ_i might be quite informative. It might, for example, give rise to point estimates for each γ_i, or to an upper and lower bound for each γ_i, with all values in the range seen, *a priori*, to be equally likely. Relatively simple extensions to the MCMC tools used previously allowed Christen and Litton (1995) to make inferences about the extra γ parameter as well as θ_1 for two examples, one relating to radiocarbon dating of a relative chronology derived from dendrochronology, and the other to calibration of data from an excavated archaeological sequence. In both cases, the inclusion of the extra prior information allowed for firmer conclusions about the calendar dates of interest than would otherwise have been possible.

Online software for undertaking Bayesian wiggle matching using this kind of approach has recently been implemented by Christen (2003) and is available at http://www.cimat.mx/Bwigg/.

A *Priori* Information about the Rate of Deposition

Christen (1994a) also developed models that allow the inclusion of prior information about the rate of deposition of dry mass in the lower level (catotelm) of peat bogs. The work is summarized in Christen et al. (1995) and suggests that we model the problem as follows. Let d_0 represent a depth datum (below which all other samples of interest will lie). θ_0 then represents the unknown age (in cal. BP) of peat at depth d_0. Peat at depth d ($d > d_0$) then has unknown calendar age θ ($\theta > \theta_0$) cal. BP. Let M represent the cumulative dry mass of material deposited between d_0 and d, let p represent the rate at which dry mass is added to the peat, and let a be proportional to the rate of decay of peat after deposition (a is assumed constant over the entire depth of peat). Adopting findings from earlier work by one of the co-authors, Christen et al. (1995) then make several suggestions for modelling peat deposition, one of which is

$$M = \frac{p}{a}\left(1 - \exp^{-a(\theta - \theta_0)}\right).$$

Utilizing this model, and building on the likelihood in Equation 1.2, Christen et al. (1995) define m_i as the cumulative dry mass at depth d_i so that

$$m_i = \frac{p}{a}\left(1 - \exp^{-a(\theta_i - \theta_0)}\right)$$

and then

$$\theta_i = \theta_0 - a^{-1}\log_e(1 - p^{-1}am_i). \tag{1.4}$$

Just as with the 'wiggle matching' model, θ_i is thus represented in terms of unknown parameters (in this case θ_0, p and a) about which we wish to learn. In much the same way, Christen et al. (1995) adopt the likelihood given in Equation 1.3, but reparameterize in terms of θ_i as defined in Equation 1.4.

Christen et al. (1995) implement this likelihood and illustrate the use of informative prior information about p and a. They apply the method to data from peat formation in two different situations, one a long-term deposition sequence and the other of much shorter duration (in which the a parameter can be ignored). In both cases the authors demonstrate that they can take account of prior information and known sources of error, thus coherently integrating a number of components that would usually be interpreted using more heuristic tools.

1.2.4 Construct Integrated Chronologies Consisting of Phases and Sequences

In this section, I consider further, very general extensions to the likelihood in Equation 1.2 that allow us to combine chronological information from different phases or periods of deposition. Imagine two phases of past human activity, one over-lying the other. Archaeologists quite often identify such sequences and will want to see them reflected in the results they get from radiocarbon calibration. I provide an illustration of this kind of relationship in Figure 1.1. A particularly dramatic example of this arises when volcanic activity, flooding or a fire destroys one house and then (several years later) people move back to use that location again. In such circumstances, we know (as a minimum) that the material found in the lower house is older than that in the upper one. If we are lucky, we will also have ordering within the stratigraphic layers in each of the houses that can inform the chronology building. Archaeological interest often focuses on the dates of construction and destruction of the two houses (and

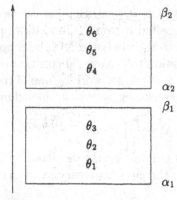

Time

Fig. 1.1. Illustration of the types of stratigraphic information and statistical parameterizations that arise when sequences of radiocarbon data can be related one to another (as in Section 1.2.4).

the length of time during which they were occupied). Unfortunately, however, we very rarely have samples suitable for radiocarbon dating that will directly date these *boundary* events. In these sorts of situations it becomes even more important to include the relative chronological prior information in a formal interpretive process.

In addition to the parameters already defined, we add phase boundary parameters (α, β) about which we have some relative chronological information, but for which there is no direct radiocarbon evidence. Assuming $j = 1, 2, \ldots, J$ groups, each bounded by a pair of unknown calendar dates (α_j, β_j) and each containing $i = 1, 2, \ldots, n_j$ dated samples, the likelihood based on $x_{i,j}$ can be expressed as

$$l(\theta_{ij}, \alpha_j, \beta_j; x_{i,j}) \propto \frac{1}{\alpha_j - \beta_j} \times \exp\left\{-\frac{(x_{i,j} - \mu(\theta_{i,j}))^2}{2\sigma_{i,j}^2}\right\},$$

and the likelihood for the whole sample, x, is given by

$$l(\theta, \alpha, \beta; x) = \prod_{j=1}^{J} \prod_{i=1}^{n_j} l(\theta_{ij}, \alpha_j, \beta_j; x_{i,j}).$$

In extending the model in this way, we needed to construct priors that represent the restrictions placed on the calendar dates by the relative chronological information, but also acknowledge that there are parts of the problem about which our prior knowledge is very vague. Conventionally, in the absence of any other prior information about $\theta_{i,j}$, we assume that it is equally likely to lie anywhere in the range α_j to β_j (but see later for reference to recent research on this topic). In the absence of any other information, we also conventionally assume *a priori* that α_j and β_j lie anywhere in the range of the calibration curve, subject to the constraint $\alpha_j > \beta_j$.

The widely used implementations of this kind of model and prior information all adopt MCMC simulation as explained in detail in Litton and Buck (1996). Depending upon the part of the calibration curve we are on, and the amount and nature of the prior information, we do sometimes have problems obtaining reliable estimates of posteriors from our MCMC. Fortunately, this is not common and we are usually able to obtain robust estimates for all parameters of interest. The parameters that we are usually most interested in are the marginals, α_j and β_j, and/or the length of time elapsed between pairs of these parameters. These can very readily be extracted from the output that arises from MCMC implementation, and are now routinely seen in archaeological reports.

1.2.5 Undertake Outlier Detection

In very many practical applications of radiocarbon dating, the presence of outliers (i.e. radiocarbon determinations that are inconsistent given the other data points, prior information and model) could grossly affect age estimates. Thus the methodology employed needs to be able to detect such data points and minimize the impact they have on the inferences we make. As a result, Christen (1994a,b) offered an important practical extension to the framework thus far outlined.

To simplify things a little, consider a single phase of the sort defined in Section 1.2.4 whose earliest and latest boundary dates are represented using α and β respectively. As before, also suppose that we have $i = 1, 2, \ldots, n$ radiocarbon determinations from within this phase. To model the possibility of having an outlier, Christen (1994a,b) suggested that the ith determination should be said to be an outlier if it needs a shift of δ_i radiocarbon years to be consistent with the rest of the samples. Formally we have

$$X_i | \theta_i, \phi_i, \delta_i \sim N\left(\mu(\theta_i) + \phi_i \delta_i, \sigma_i^2\right),$$

where

$$\phi_i = \begin{cases} 1 & \text{if the } i\text{th determination needs a shift} \\ 0 & \text{otherwise.} \end{cases}$$

Thus an error in the radiocarbon dating process of the ith determination will be modelled by a shift, δ_i, in the observed value from the expected true radiocarbon age $\mu(\theta_i)$. The prior uncertainty concerning whether or not a shift is needed is measured by the prior probability we associate with $\phi_i = 1$ and $\phi_i = 0$ respectively. Christen (1994a,b) assumed that whether or not the ith determination needs a shift (and the size of the shift where required) is independent of the other determinations and of α and β. As a result, the likelihood he derives can be summarized as

$$l(\boldsymbol{\theta}, \boldsymbol{\delta}, \boldsymbol{\phi}, \alpha, \beta; \boldsymbol{x}) \propto (\alpha - \beta)^{-n} \prod_{i=1}^{n} z_i,$$

where
$$z_i = \exp\left\{-\frac{(x_i - \delta_i\phi_i - \mu(\theta_i))^2}{2\sigma_i^2}\right\}.$$

As for previous models, our prior information about the boundary dates is typically vague and so

$$p(\alpha, \beta) \propto 1 \quad \text{for } \beta < \alpha.$$

For most radiocarbon samples, our prior information about how big a shift to expect is also quite vague. As a result, a simple form for the prior is

$$\delta_i \sim N(0, a\sigma_i^2)$$

where a is a suitable constant. For example, in the *BCal* implementation of this problem (Buck et al. 1999), $a = 4$. In simple terms, this means that *a priori* a sample is seen as an outlier if it needs a shift of more than twice the size of its reported laboratory error. When it comes to prior information about the probability of a specific sample being an outlier, this is quite often more informative. We represent this prior probability as q_i, so that the prior probability that $\phi_i = 1$ is q_i and the prior probability that $\phi_i = 0$ is $(1 - q_i)$. In other words our prior assessment that the ith determination needs a shift is q_i and that it does not need a shift is $1 - q_i$. Unless we have case-specific expert knowledge to the contrary, Christen (1994a,b) suggests that we see $q_i = 0.1$ as representing vague prior information.

This approach to outlier detection scales very nicely to situations where we have multiple phases, can readily be implemented using MCMC (for details see Christen 1994a), and forms part of the widely used software known as *BCal* (http://bcal.shef.ac.uk/). It allows users to

- assume *a priori* that all samples are reliable (and hence set $P(\phi_i = 1) = 0$); or
- to suppose that all samples have a fixed prior probability of being an outlier (say 0.1) and set q_i accordingly; or
- to assign a range of prior outlier probabilities, with higher probabilities associated with samples that are likely to have problems (such as contamination of samples prior to dating).

In all cases where the probability that $\phi_i = 1$ is greater than zero, the *a posteriori* probability that i needs a shift to come in-line with the other samples in the group is computed as part of the MCMC analysis. This *a posteriori* probability is then used to weight the impact that sample i has on the calendar date estimates obtained. When there is a high *a posteriori* probability that sample i is an outlier, that sample has a reduced impact on the calendar date estimates we obtain.

1.3 Existing and Desirable Extensions

1.3.1 Sample Selection

One of the powerful features of Bayesian methods has not received as much attention from chronology builders as one might expect. This is the fact that we can use the paradigm to allow us to undertake 'what if' experiments before undertaking expensive or difficult further research. We can simulate new data with particular properties to see what effect information of this type would have on the inferences we would make. Alternatively, we can vary the way in which we model or represent our prior information to see how such changes will alter the inferences we make. While these features of Bayesian research are well known amongst statisticians, only a limited range are implemented in the standard Bayesian chronology building software and are thus not routinely in use.

A particularly useful existing tool is that illustrated by Bayliss and Bronk Ramsey (Chapter 2) which is provided by *OxCal*. It allows users to simulate radiocarbon determinations associated with particular calendar dates or from particular parts of models (via the R_Simulate tool) and has proved invaluable to those planning large dating programmes.

By way of illustration of other sample selection methods, Buck and Christen (1998) sought to find ways to help archaeologists who had already undertaken a chronology building project and were now interested in selecting samples from those held in store. We were seeking to answer questions of the form: "Which samples should I send for dating next in order to optimize the quality of the chronology I build?" The approach we chose was a risk analysis one which allowed us to trade gain in knowledge against the cost incurred in obtaining the new data. The basis of the approach is that for each set of samples sent for dating we obtain a predicted gain in knowledge while at the same time incurring a particular cost for the work. The optimum selection of samples to send for dating thus corresponds to the equilibrium point at which gain in knowledge is maximal and cost is minimal. In order to use tools of this type, we need to:

- establish exactly what we are attempting to learn about,
- find a way to measure the level of our current knowledge (or ignorance), and
- devise a function to evaluate the cost of a particular selection of samples that we might send for dating.

In our paper (Buck and Christen 1998) we formalized these as follows.

- We defined what we were trying to learn about in terms of the date of one or more events identified during excavation.
- For our measure of current knowledge, we chose the sum of the variances for all the parameters of interest.

- In devising a function to evaluate the cost of a particular selection of samples there were a number of things we felt we had to take into account. Clearly, we needed to include the financial cost of purchasing radiocarbon determinations from the laboratory. We felt that it was also important to chose a measure that took account of the costs already incurred in attaining the current state of knowledge and the costs associated with the time taken to prepare samples to send them for dating (these could include both field and post-processing costs). The details of the function we chose are in the original paper; here it is sufficient to note that it is flexible and could be substituted with something completely different if others wished to focus on different (possibly non-monetary) types of cost.

With the three factors thus defined, we implemented a set of simulation tools to allow us to estimate the level of knowledge that might be attained by sending different sets of samples for dating. The simulation code needed to do this is not that complex, but it does require large amounts of CPU time to run. The specific computing power needed is, of course, highly dependent upon the precise problem under investigation, but for the example we published in 1998 this took several hundred hours of CPU time. We were able to shorten the real time taken to around 80 hours by adopting a parallel processing approach using several workstations simultaneously, but this approach would not be available to very many archæologists.

Given that the method we developed is flexible, scalable and (potentially) suitable for use with a range of dating techniques, I feel that it could help enormously with planning dating programmes if we could make it more widely available. Since we did that initial work there have been quite major advances in the power and sophistication of simulation techniques. In addition, there have been considerable improvements in computing power. It may thus be worth investing a bit more effort in trying to implement a faster range of algorithms based on our initial theory.

1.3.2 Model Choice

Another aspect of applied Bayesian methods, which is yet to be widely adopted for chronology building, is formal model selection or model choice. This topic has been receiving increased interest in many applied disciplines recently with both relative and absolute chronology building standing to benefit. Since the potential of formal model choice is covered in some detail by Sahu (Chapter 5), here I simply summarize two recent projects that are particularly relevant to other material in this chapter.

Model Choice for Relative Chronology Building

In a recent paper, Buck and Sahu (2000) demonstrated the use of formal model choice tools in the context of relative chronology building via seriation. This paper describes implementations of two plausible models for seriation; the archaeologically well-known Robinson–Kendall model (Robinson 1951; Kendall 1970, 1971; Buck and Litton 1991) and a model-based approach to the widely used exploratory tool known as correspondence analysis (Baxter 1994, chap. 5). Since seriation is a chronology building tool that has been tailored to the needs of archaeologists and is not particularly relevant to researchers in other disciplines, I will not focus on the details of the two models here. What is more generally relevant, is how we chose between them when making inferences based on a particular data set.

We developed a decision-theoretic approach, based on loss functions. The technical details of the approach are given in the original paper; here I provide an intuitive summary. Let $\mathbf{n}_{\text{obs}}$ represent observed data with individual data points denoted by $n_{ij,\text{obs}}$, $i = 1, 2, \ldots, I$ and $j = 1, 2, \ldots, J$. In seriation, this might be the count of a particular burial offering, j, in an excavated human grave, i. Similarly, let $\mathbf{n}_{\text{rep}}$ with components $n_{ij,\text{rep}}$ (where rep is an abbreviation for replicate) denote a set of observables predicted by a particular model. The current model is then a good fit to the observed data if $\mathbf{n}_{\text{rep}}$ replicated $\mathbf{n}_{\text{obs}}$ well. So, we need a way to measure the quality of the fit. In Buck and Sahu (2000) we adopted a simple loss function for measuring the divergence between observations and replicates. It took the form

$$L(\mathbf{n}_{\text{rep}}, \mathbf{n}_{\text{obs}}) = 2 \left(\sum_{ij} n_{ij,\text{obs}} \log \frac{n_{ij,\text{obs}}}{n_{ij,\text{rep}}} \right)$$

which leads to high values if the data predicted by the model, $n_{ij,\text{rep}}$, are **not** close to the observed data, $n_{ij,\text{obs}}$. The best model among a set of models is then the one for which the expected value of the loss function is minimum.

We demonstrated the use of this approach for two well-known small data sets, finding in both cases that our Bayesian implementation of the Robinson–Kendall model provides a considerably better fit to the data than our correspondence analysis model (Buck and Sahu 2000). As with much research in applied Bayesian statistics, the algorithms we developed are computationally intensive and require some tailoring from one data set to another. As a result these tools are not yet routinely available for use by archaeologists. If there is sufficient demand, however, this is another area in which further work on developing and generalizing computer code might be worthwhile.

Model Choice for Absolute Chronology Building

Nicholls and Jones (2001) have implemented a rather different approach to model choice than the one outlined above. They use Bayes factors to help select between competing models for absolute chronologies based on radiocarbon determinations (for a general explanation of Bayes factors and details of their use in Bayesian model choice see Berger 1999). Again, I won't try to give any details relating to Bayes factors, but I can give an intuitive explanation. Suppose we have two plausible models that might be used to represent a particular chronology about which we are attempting to learn. We represent these models as M_1, corresponding to a chronological problem formulated as $f^{(1)}$, and M_2 for $f^{(2)}$. For model M_k, $k = 1, 2$, let $P(M_k \mid x)$ represent the *a posteriori* preference for model k given the current data, x, and let $P(M_k)$ give the prior preference for model k (before the data are observed). The Bayes factor $B(2,1)$ for M_2 against M_1 is given by the ratio of the *a posteriori* probabilities for the two models,

$$B(2,1) \equiv P(M_2 \mid y)/P(M_1 \mid y).$$

Now, if we believe *a priori* that both models are equally likely to represent our chronology reliably, then $P(M_1) = P(M_2)$. Thus, the component of the posterior that comes from the prior is the same for both models and so the ratio of the *a posteriori* probabilities for the two models equals the ratio of their likelihoods. In other words, $B(2,1) = P(y \mid M_2)/P(y \mid M_1)$.

Nicholls and Jones (2001) compute the model likelihoods using the estimators given in Meng and Wong (1996) and find them to be fairly straightforward to implement for their case study. Since there are quite a number of ways to formulate Bayesian model choice, however, and Nicholls and Jones and others including Sahu (Chapter 5) are currently working on this problem, it will be a little while before these tools become widely available in general purpose software.

1.3.3 Alternative Prior Specifications

The formalization and careful representation of prior information is essential to the robust and reliable use of Bayesian methods. Very often in chronology building we have only limited prior knowledge about the likely dates of events of interest. In these situations, Bayesians seek ways formally to represent our *a priori* lack of knowledge – we call this non-informative or vague prior information. In all early implementations of Bayesian chronology building non-informative priors were constructed by assuming that the date of each event was equally likely to lie anywhere on an appropriate time interval (in the case of radiocarbon dating, anywhere on the calibration curve) provided that any stratigraphic or other ordering constraints were obeyed. In this way, *a priori*, the date of each event is taken to be uniform on a fixed (and often arbitrary) interval.

Recent research (Nicholls and Jones 2001) has shown that by constructing our priors in this way we are in fact biasing towards long (rather than short) date ranges between pairs of parameters. This is particularly problematic in situations where we have relatively little chronometric information. This prompted Nicholls and Jones (2001) to propose an alternative formulation for non-informative priors in which the *difference* between the earliest and latest dates (i.e. $\beta_j - \alpha_j$) is taken to have a uniform distribution (for an illustration of the form of these priors see Figure 5.1 in this volume). Nicholls and Jones (2001) implement both the traditional non-informative prior and their new one and illustrate some of the properties of their new prior formulation using an example from New Zealand prehistory. While many researchers are still using the more traditional non-informative prior, the one suggested by Nicholls and Jones is already an integral part of the radiocarbon calibration software known as *OxCal* (Bronk Ramsey 1995).

There are currently a number of research projects underway in which different types of informative and non-informative prior specification are being considered. Along with model choice work, this is one of the most interesting areas of current work in chronology building and one that has potential to alter dramatically the way we formalize our ideas.

1.3.4 Modelling the Radiocarbon Calibration Curve

Traditionally, both Bayesian and classical calibration tools have used the radiocarbon calibration data in their piece-wise linear form, joining the means of the calibration data points one to another, or by proposing spline functions through the calibration data. There is potential here for more formal modelling, however, and this is another active area of research in Bayesian chronology building. As part of his PhD work Christen (1994a) suggested that we should consider a formal model for incorporating the errors on the calibration data as part of the calibration process. His model is simple, adding a formalization of the error to the existing piece-wise linear fit. It is implemented in *BCal* (Buck et al. 1999), and allows researchers to formally include this further source of error in the calibration process. This model should be seen as a first step towards modelling the calibration data since it does not provide us with a way to represent any of the processes that gave rise to the radiocarbon data we observe.

Gómez Portugal Aguilar (2001) has attempted to address this problem in her recently completed PhD work. She suggests a new, fully Bayesian approach to utilizing the curve data in which the radiocarbon calibration curve is seen as an unknown function that needs to be estimated (in the presence of the errors on the curve data themselves). Gómez Portugal Aguilar makes inferences about the whole calibration function, based on the information provided by the calibration data, and incorporating prior beliefs about important features of the curve, such as smoothness and differentiability. She models the radiocarbon calibration curve using a Gaussian process prior distribution on the space of all possible functions, specified according to clearly stated prior information regarding the underlying process of radiocarbon generation. The result is an estimate of the curve which takes account of a wide range of sources of error (including those in the calibration data themselves). One particular feature that arises from this approach is that the estimate of the calibration curve obtained (i.e. the *a posteriori* mean) does not interpolate (go through the means of) the data points. Moreover, the resulting variance values for the calibration curve seem more realistic than those resulting from other approaches. A preliminary report on this work is given in Gómez Portugal Aguilar et al. (2002) and a more detailed paper is in preparation (Gómez Portugal Aguilar et al. in preparation). Although considerable debate is still needed as to the pros and cons of routinely adopting one particular smooth form for the calibration curve, this is an important step towards more realistic modelling of the problem and future work in this area will be worth following with interest. This is particularly true now that similar approaches are being adopted for modelling the archaeomagnetic calibration curve (see Chapter 3).

'Hybrid' Radiocarbon Samples

For simplicity, many people interpreting radiocarbon determinations assume that the samples they send for radiocarbon dating arise from organisms that metabolized material from either an entirely marine or an entirely terrestrial source. This is the most straightforward approach to take because the radioactive isotope of carbon (^{14}C) mixes quite differently in the atmosphere and the oceans and so two different calibration curves exist. Quite often, however, dietary evidence and/or ^{13}C/^{12}C ratios suggest that organisms metabolized material from a mixture of the two environments. This is particularly relevant at the moment in archaeology as quite large funds are being spent on dietary analysis of human material from island and coastal environments. Examples of such projects include: *A question of time and place: palaeodietary and dating investigations in the Mesolithic and Neolithic of Brittany* funded by NERC, and *Stable isotope investigations in the Mesolithic of coastal Wales and SW England* funded by the British Academy (Board of Celtic Studies) and by NERC via the ORADS dating programme.

Such work involves stable isotope analysis of human bone, as well as radiocarbon dating, and gives rise to estimates of the proportion of marine and terrestrial material in the diet of the individual concerned. In order to make use of the dietary information, in the interpretation of radiocarbon data, some researchers are adopting very simple algorithms to undertake what is becoming known as 'hybrid' calibration (implemented, for example, in *OxCal* and described at http://www.rlaha.ox.ac.uk/oxcal/math_ca.htm#mix_curves).

Given the complex nature of the models needed for integrated interpretation of radiocarbon data, and the results of some unpublished work (Buck and Christen 2000; Rouse 2002), I believe that such simple algorithms deserve further investigation before they become more widely adopted. We need to look in detail at modelling two important issues related to this problem. The first is the quantification and interpretation of the errors associated with the estimates of dietary proportions. The second is the precise manner in which we combine marine and terrestrial calibration data in the same interpretive process.

1.3.5 Non-Archaeological Chronology Building

As is clear from several other chapters in this volume, archaeology is not the only discipline in which chronology building is important. Geologists, for example, use chronometric dating evidence to try to build up a picture of the absolute chronology of the part of the Earth that interests them. Although not the most common chronometric dating technique in geology, radiocarbon dating is routinely used for work on more recent material such as those derived from (or associated with) recent volcanic activity. Despite the success of Bayesian methods for the interpretation of radiocarbon determinations in archaeology, until recently there has been very little interest in Bayesian methods amongst geologists. In fact, many geologists do not calibrate their radiocarbon determinations, they simply treat radiocarbon ages as if they were calendar dates. Traditionally, this has been held to be acceptable because of the coarse scale at which geologists often build chronologies, however as they increasingly seek to pin down the dates of single geological events, they will encounter all the same problems that archaeologists have had in the past (not least of which is that the wiggles in the calibration curve result in inversions of the orders of events between the radiocarbon and the calendar scale). I feel sure that we will see more on this in the geological literature very soon, not least because of recommendations from the INTIMATE group (Lowe et al. 2001) which, among other things suggests that more attention should be paid to Bayesian techniques for calibrating stratigraphic sequences.

As a step towards greater acceptance, colleagues and I have reworked data relating to the Kaharoa eruption of Mount Tarawera in the North Island of New Zealand (Buck et al. 2003). Precise dating of this eruption is important not just to geologists, but to archaeologists too. Since the eruption was explosive, it showered an enormous area of the surrounding landscape with fine ash. When identified in geological sequences, such layers are known as tephra. For more on the use of tephra in chronology building see Dugmore et al. (Chapter 8). Here it is sufficient to note that the physical and chemical make-up of the ash in tephra can often be used to uniquely link it to a particular eruptive episode. Since tephra layers are readily identifiable in excavated deposits both on archaeological sites and in geological sequences, if they could be reliably dated they would provide a most useful chronological marker. Unfortunately, the dating of the Kaharoa tephra is complicated by the fact that this eruption happened at a time period where there is a large inversion in the radiocarbon calibration data and so all calibrated radiocarbon date estimates exhibit multi-modality and wide ranges. Thus, any conventional calibration, which fails to make use of prior information, will typically not allow very close dating of the event of interest. By adopting a selection of the Bayesian models outlined in Section 1.2 (including outlier detection), we have been able to help improve the interpretation of radiocarbon dates associated with the Kaharoa eruption. Alongside this, there has been recent work to provide a 'wiggle match' date for this event and it thus seems likely that we will soon have a definitive date.

1.4 Other Chronometric Dating Techniques

Although radiocarbon dating is still the single most widely used chronometric dating technique in archaeology, there are a number of others that are important to archaeologists and researchers in other disciplines and it would be really useful if we could include a wider range of chronometric dating evidence in the Bayesian chronology building process. There has already been considerable research into Bayesian methods for dendrochronology (tree-ring dating) (Litton and Zainodin 1991; Zainodin 1988). This work was, however, undertaken in the late 1980s and has not been widely discussed since. Millard (2002) has revived interest in this area by suggesting an approach for sapwood estimates and felling dates.

Alongside work on dendrochronology, several other researchers around the world are seeking Bayesian methods to aid in the interpretation of a range of other chronometric dating techniques. Jones (2002) reports on a formalization of a suitable likelihood for obsidian hydration dating, Millard is also currently working on suitable likelihoods for U-series, luminescence and ESR dating (see Chapter 11), and Lanos (in preparation) is working on a Bayesian approach to the calibration and interpretation of archaeomagnetic data (see also Chapter 3).

1.5 Spatio-Temporal Modelling

Although current Bayesian models for chronology building all focus on temporal data interpretation from a single spatial location, in practice we rarely want to interpret the chronology of one site in isolation from others in the same landscape. As a result another active area of research in Bayesian chronology building is the search for robust and elegant ways to add a spatial component to existing models. We have already identified two possible ways forward (summarized below), but this is an active area of research and I also expect other ideas to come to the fore.

1.5.1 Using Existing Models

In recent work Blackwell and Buck (2003) have reanalysed data relating to the post-glacial human recolonization of NW Europe, using relatively simple Bayesian data interpretation tools. The interpretation of these data has been debated as either indicating wave-of-advance colonization (Housley et al. 1997) or showing very little evidence for movement of peoples across Europe during the last deglaciation (Blockley et al. 2000). However, due to the heuristic nature of the approaches that were adopted there is no formal way to compare these methods and results and, thus, no obvious means to resolve the debate. As a result, Blackwell and Buck (2003) have modelled radiocarbon dated samples in each region as deposited between a start date (α) and an end date (β). We acknowledged that it was unlikely we would ever gain direct radiocarbon evidence about the date of earliest colonization in a region, and thus estimated it using the method summarized in Section 1.2.4, assuming constant deposition rates between α and β (this time across an entire geographical region rather than at a single archaeological site). Given statistical estimates for the date of initial colonization of each region, we then computed the probability of all possible orders of recolonization. Despite the considerable efforts expended in collecting and dating samples relating to this recolonization, all date estimates arising from the pilot study were associated with rather large uncertainty intervals and thus drawing firm conclusions on the ordering was most difficult. This is because the models used successfully represent the uncertainty present in radiocarbon dating. To draw firmer conclusions requires more chronological data and/or more sophisticated models that include the spatial structure believed to be present, but which is currently unrepresented. In our paper we make some suggestions for the kinds of models that might help in future research of this kind.

1.5.2 Point Process Models

In the second project, Buck et al. (in preparation) have been devising models aimed at aiding the interpretation of material from neighbouring, but distinct, excavations. We have in mind the Shag River Mouth area in New Zealand which is the most extensively dated coastal landscape that is known to have been exploited by early human settlers of New Zealand (Anderson et al. 1996a). Interpretation of the chronology of this region is influential in understanding the timing of initial human settlement of the whole of New Zealand. Until the early 1990s, there was little in the way of chronometric data from the region, but the presence of deep stratigraphy (in places up to 2.5 m, comprising many distinct layers) and other archaeological indicators suggest that the occupation span of the site was long, possibly centuries. In a re-evaluation of the site alongside a review of recent radiocarbon dating evidence, Anderson et al. (1996b) have suggested that this earlier interpretation may be incorrect and that deposition could have taken place over as little as 20 to 40 years.

Building on some earlier work by Nicholls and Jones (2001), Buck et al. (in preparation) hope to make a further contribution to this debate. Since the data from the Shag River Mouth are both spatial and temporal in nature, integration of data from the entire landscape rather than from one stratigraphic sequence at a time seems desirable. Towards this end, we have identified an integrated modelling approach which seems likely to lead to a scalable solution. By modelling excavated material at different locations as having been deposited via linked Poisson point processes, we obtain models that are truly spatio-temporal in nature, but also have many of the same properties as the models currently used in Bayesian radiocarbon calibration. Similar models are used by Rodriguez-Iturbe et al. (1988) to model rain as falling according to a hierarchical point process of randomly varying intensities. In our study, events likewise fall according to point processes arranged in a hierarchy. Major, and typically undated, structuring events (such as the founding or destruction of settlements) are used to act as bounding events for lower-level processes, which generate more localized events (such as food preparation). The model connects radiocarbon dates associated with localized events to the higher-level events, which are of real interest.

In this way, we are able to allow chronometric data from one spatial location to help us learn about the chronology of the use of a whole landscape even when individual excavated layers cannot be linked one to another. Such a model is clearly extensible and has potential applications both for the interpretation of data from small, regional studies and for large-scale colonization and recolonization projects covering entire countries or continents.

1.5.3 Future Work on Spatio-Temporal Data Interpretation

Both of the ideas outlined above indicate that a formal, clear and coherent model of spatio-temporal information allows previously unobtainable probabilistic inferences to be made. They have also identified a number of important research topics that still need tackling before such tools can be used routinely. In general terms this research leads me to believe that there are two different, but closely related, ways to represent spatio-temporal movement of people, ideas and technologies. The first is simply to use a statistical model for how α and β vary with geographical location. Such models allow us to represent the idea that, although we may not know a priori which locations were recolonized first, all other things being equal, we do expect dates for (geographically) neighbouring locations to be more similar than those for locations a long way apart. This is natural, simply because such movements were spatio-temporal, rather than something occurring independently at different locations. The second way is to formulate an explicit, mechanistic model of the movements that took place. Such models have been suggested in the past and are well known, e.g. those of Ammerman and Cavalli-Sforza (1971, 1984) or Steele et al. (1998). The models implemented by these authors do not, however, have the stochastic components needed to incorporate archaeometric data. They will need considerable tailoring to allow coherent handling of a range of chronometric data and prior spatio-temporal information.

1.6 Summary

In summary, Bayesian methods are already seen by many as an essential tool to aid in formal chronology building in archaeology. At present, most researchers use packages like *OxCal* and *BCal* to make use of such tools and typically see them as radiocarbon calibration tools (indeed both are described as such on their own WWW welcome pages). On reflection, however, I think that it is clear that these packages offer more than just calibration, they are modest Bayesian chronological data interpretation environments. Given this observation, and the fact that the current tools are built on a sound foundation of flexible and scalable theory, I think that we are in a good position to move towards fully integrated tools for Bayesian chronology building. All of the current and planned research projects outlined above will contribute to the extension of the framework in one way or another. Since such work is motivated by a desire to provide practical solutions to real, current and pressing issues associated with chronology building, I feel sure that we can look forward to many more years of fast moving, productive and practical research in Bayesian chronology building.

References

Ammerman, A. J. and Cavalli-Sforza, L. L. (1971). Measurement of the rate of spread of early farming in Europe. *Man*, **6**, 674–688.

Ammerman, A. J. and Cavalli-Sforza, L. L. (1984). *The Neolithic transition and the genetics of populations in Europe*. Princeton University Press, Princeton.

Anderson, A., Allingham, B. and Smith, I. (1996a). *Shag River Mouth: the archaeology of an early southern Maori village*. ANH Publications, RSPAS, Australian National University, Canberra.

Anderson, A., Smith, I. and Higham, T. F. G. (1996b). Radiocarbon chronology. In A. Anderson, B. Allingham and I. Smith (eds.), *Shag River Mouth: the archaeology of an early southern Maori village*, ANH Publications, RSPAS, Australian National University, Canberra, chap. 7, 60–69.

Baxter, M. J. (1994). *Exploratory multivariate analysis in archaeology*. Edinburgh University Press, Edinburgh.

Berger, J. O. (1999). Bayes factors. In S. Kotz, C. B. Reed and D. L. Banks (eds.), *Encyclopedia of statistical sciences, update volume 3*, Wiley Inter-Science, New York, 20–29.

Blackwell, P. G. and Buck, C. E. (2003). The Late Glacial human reoccupation of north western Europe: new approaches to space–time modelling. *Antiquity*, **77**, 232–240.

Blockley, S. P. E., Donahue, R. E. and Pollard, A. M. (2000). Radiocarbon calibration and late glacial occupation in northwestern Europe. *Antiquity*, **74**, 112–119.

Bronk Ramsey, C. (1995). Radiocarbon calibration and analysis of stratigraphy: the OxCal program. *Radiocarbon*, **37**, 425–430.

Buck, C. E., Cavanagh, W. G. and Litton, C. D. (1996). *Bayesian approach to interpreting archaeological data*. John Wiley, Chichester.

Buck, C. E. and Christen, J. A. (1998). A novel approach to selecting samples for radiocarbon dating. *Journal of Archaeological Science*, **25**, 303–310.

Buck, C. E. and Christen, J. A. (2000). A Bayesian calibration of radiocarbon determinations from the Pearl Harbor Fishponds, Hawai'i. In J. S. Athens (ed.), *Ancient Hawaiian fishponds of Pearl Harbor; archaeological studies on US Navy Lands, Hawai'i*, Department of Defense Legacy Resource Management Program, Project Number 1729, Part C.

Buck, C. E., Christen, J. A. and James, G. N. (1999). BCal: an on-line Bayesian radiocarbon calibration tool. *Internet Archaeology*, **7**. URL http://intarch.ac.uk/journal/issue7/buck/

Buck, C. E., Christen, J. A., Kenworthy, J. B. and Litton, C. D. (1994a). Estimating the duration of archaeological activity using ^{14}C determinations. *Oxford Journal of Archaeology*, **13**, 229–240.

Buck, C. E., Higham, T. F. G. and Lowe, D. J. (2003). Bayesian tools for tephrochronology. *Holocene*, **13**, in press.

Buck, C. E., Jones, M. and Nicholls, G. (in preparation). A Bayesian framework for the interpretation of spatio-temporal data.

Buck, C. E., Kenworthy, J. B., Litton, C. D. and Smith, A. F. M. (1991). Combining archaeological and radiocarbon information: a Bayesian approach to calibration. *Antiquity*, **65**, 808–821.

Buck, C. E. and Litton, C. D. (1991). A computational Bayes approach to some common archaeological problems. In K. Lockyear and S. P. Q. Rahtz (eds.), *Computer applications and quantitative methods in archaeology, 1990*. British Archaeological Reports, Oxford, International Series, **S565**, 93–100.

Buck, C. E. and Litton, C. D. (1995). The radiocarbon chronology: further consideration of the Danebury dataset. In B. Cunliffe (ed.), *Danebury: an Iron Age hillfort in Hampshire, Volume 6, a hillfort community in Hampshire*, Council for British Archaeology, Report Number 102, 130–136.

Buck, C. E., Litton, C. D. and Scott, E. M. (1994b). Making the most of radiocarbon dating: some statistical considerations. *Antiquity*, **68**, 252–263.

Buck, C. E., Litton, C. D. and Shennan, S. J. (1994c). A case study in combining radiocarbon and archaeological information: the early Bronze Age settlement of St. Veit-Klinglberg, Land Salzburg, Austria. *Germania*, **72**, 427–447.

Buck, C. E., Litton, C. D. and Smith, A. F. M. (1992). Calibration of radiocarbon results pertaining to related archaeological events. *Journal of Archaeological Science*, **19**, 497–512.

Buck, C. E. and Sahu, S. K. (2000). Bayesian models for relative archaeological chronology building. *Applied Statistics*, **49**, 423–440.

Christen, J. A. (1994a). *Bayesian interpretation of* ^{14}C *results*. Ph.D. thesis, University of Nottingham, Nottingham, UK.

Christen, J. A. (1994b). Summarizing a set of radiocarbon determinations: a robust approach. *Applied Statistics*, **43**, 489–503.

Christen, J. A. (2003). Bwigg: an Internet facility for Bayesian radiocarbon wiggle-matching. *Internet Archaeology*, **13**.
URL http://intarch.ac.uk/journal/issue13/christen_index.html

Christen, J. A., Clymo, R. S. and Litton, C. D. (1995). A Bayesian approach to the use of ^{14}C dates in the estimation of the age of peat. *Radiocarbon*, **37**, 431–442.

Christen, J. A. and Litton, C. D. (1995). A Bayesian approach to wiggle-matching. *Journal of Archaeological Science*, **22**, 719–725.

Gilks, W., Richardson, S. and Spiegelhalter, D. (eds.) (1996). *Markov chain Monte Carlo in practice*. Chapman and Hall, London.

Gómez Portugal Aguilar, D. (2001). *Bayesian modelling of the radiocarbon calibration curve*. Ph.D. thesis, University of Sheffield, Sheffield, UK.

Gómez Portugal Aguilar, D., Buck, C. E., Litton, C. D. and O'Hagan, A. (in preparation). Bayesian non-parametric estimation of the calibration curve for radiocarbon dating.

Gómez Portugal Aguilar, D., Litton, C. D. and O'Hagan, A. (2002). Novel statistical model for a piece-wise linear radiocarbon calibration curve. *Radiocarbon*, **44**, 195–212.

Housley, R. A., Gamble, C. S., Street, M. and Pettitt, P. (1997). Radiocarbon evidence for the Lateglacial human recolonisation of Northern Europe. *Proceedings of the Prehistoric Society*, **63**, 25–54.

Jones, M. D. (2002). *A brief prehistory of time*. Ph.D. thesis, University of Auckland, Auckland, New Zealand.

Kendall, D. G. (1970). A mathematical approach to seriation. *Philosophical Transactions of the Royal Society Series A*, **269**, 125–135.

Kendall, D. G. (1971). Seriation from abundance matrices. In F. R. Hodson, D. G. Kendall and P. Tautu (eds.), *Mathematics in the archaeological and historical sciences*, Edinburgh University Press, Edinburgh, 215–252.

Lanos, P. (in preparation). Archaeomagnetic reference curves, Part II: Bayesian approach to smoothing multivariate and spherical time series carrying errors of both date and measurement.

Litton, C. D. and Buck, C. E. (1996). An archaeological example: radiocarbon dating. In S. R. W. Gilks and D. Spiegelhalter (eds.), *Markov chain Monte Carlo in practice*, Chapman and Hall, London, 465–480.

Litton, C. D. and Zainodin, H. J. (1991). Statistical models of dendrochronology. *Journal of Archaeological Science*, **18**, 429–440.

Lowe, J. J., Hoek, W. Z. and INTIMATE Group (2001). Inter-regional correlation of palaeoclimatic records for the last glacial–interglacial transition: a protocol for improved precision recommended by the INTIMATE project group. *Quaternary Science Reviews*, **20**, 1175–1187.

Meng, X.-L. and Wong, W. H. (1996). Simulating ratios of normalising constants via a simple identity: a theoretical exploration. *Statistica Sinica*, **6**, 831–860.

Millard, A. R. (2002). A Bayesian approach to sapwood estimates and felling dates in dendrochronology. *Archaeometry*, **44**, 137–144.

Naylor, J. C. and Smith, A. F. M. (1988). An archaeological inference problem. *Journal of the American Statistical Association*, **83**, 588–595.

Nicholls, G. and Jones, M. (2001). Radiocarbon dating with temporal order constraints. *Applied Statistics*, **50**, 503–521.

Robinson, W. S. (1951). A method for chronologically ordering archaeological deposits. *American Antiquity*, **16**, 293–301.

Rodriguez-Iturbe, I., Cox, D. R. and Isham, V. (1988). A point process model for rainfall: further developments. *Proceedings of the Royal Society of London Series A*, **417**, 283–298.

Rouse, V. (2002). *Dating archaeological samples of 'hybrid' terrestrial and marine origin.* Master's thesis, Department of Statistics, University of Sheffield, Sheffield, UK.

Steele, J., Adams, J. and Sluckin, T. (1998). Modelling Paleoindian dispersals (paleoecology and human populations). *World Archaeology*, **30**, 286–305.

Stuiver, M., Reimer, P. J., Bard, E., Beck, J. W., Burr, G. S., Hughen, K. A., Kromer, B., McCormac, F. G., Plicht, J. V. D. and Spurk, M. (1998). IntCal98 radiocarbon age calibration, 24,000-0 cal BP. *Radiocarbon*, **40**, 1041–1083.

Zainodin, H. J. (1988). *Statistical models and techniques for dendrochronology.* Ph.D. thesis, University of Nottingham, Nottingham, UK.

2

Pragmatic Bayesians: a Decade of Integrating Radiocarbon Dates into Chronological Models

Alex Bayliss and Christopher Bronk Ramsey

Summary. This chapter is an account of the experiences of the two authors in routinely applying Bayesian statistics to sets of radiocarbon dates. Whereas Buck (Chapter 1) gives an account of the development of the mathematical models, this chapter is concerned with their implementation and routine use. Since the early 1990s radiocarbon data from a large number of archaeological sites have been analysed using Bayesian methods and this chapter provides an overview of the subject from an archaeological perspective. Using examples largely taken from English Heritage projects, the chapter demonstrates the usefulness of the technique, some of the pitfalls and problems encountered and areas for future research. The problems that arise when trying to apply rigorous mathematical methods to very complex models are discussed in relation to the developments of the computer program *OxCal*. Other perspectives on archaeological chronology building are given by Holst (Chapter 6) and Cichocki et al. (Chapter 4).

2.1 Introduction

As a physicist working in radiocarbon dating, Bronk Ramsey's first foray into Bayesian statistics was while working on the dating of the Ice Man (Hedges et al. 1992; Rom et al. 1999). The dates obtained from the skin and the bone did not quite agree – the dates on the skin being significantly younger. A look at the calibration curve showed that there was a steep part of the calibration curve at this point. Given that the carbon turnover rate in bone for an adult was thought to be about 20 years longer than that in skin, it was possible that this was the explanation. A method was needed to combine the calibrations of two dates thought to be about 20 years apart. Code was written to generate the probability distributions, and a suite of algorithms produced to perform mathematical manipulations on them. This was the basis for *OxCal*.

By this time, Caitlin Buck and Cliff Litton had published work on the application of Bayesian statistics to archaeological dating (Buck et al. 1991) and, following very useful meetings with them, *OxCal* was developed to try to reproduce their results – an exercise which proved useful to all concerned.

Since then development has largely been driven by the types of problems encountered, and by a wish to make *OxCal* a useful tool for those wishing to try this type of analysis. Useful advice has come from many people – in particular Geoff Nichols and Peter Steier, and of course ideas from the original work of Caitlin Buck and those working with her. There are now over 200 registered users (and probably many more unregistered ones as the program is freely distributed) in 22 countries – though it must be said the majority of these are probably only using the program's calibration and display facilities!

As the Scientific Dating Co-ordinator of English Heritage, Bayliss is responsible for ensuring that radiocarbon dating is applied to maximum effect on archaeological projects funded by English Heritage, and that this is done in the most cost-effective manner possible. Coming from a background in quantitative archaeology, the potential of the Bayesian approach for using archaeological information to refine radiocarbon dating in a explicit manner was obvious from the appearance of the first papers. The only obstacle was the pragmatic difficulties of applying the technique routinely, with a consumer base which was (and is!) almost entirely innumerate.

Since the introduction of *OxCal*, almost 2000 radiocarbon determinations have been commissioned and analysed from more than 180 projects funded by English Heritage. This represents an investment of well over £600,000 of tax-payers' money. Of this, 66% has been spent in a framework of Bayesian sample selection, although this is not distributed evenly between different types of project (Figure 2.1). For excavations, around 85% of radiocarbon samples have been selected in this way, but for survey projects less than 11% use this methodology.

A rough idea of the effectiveness of the approach can be found by comparing the number of years spanned by the posterior density estimates from the chronological models used, with the number of years spanned by the simple calibrated dates of the same measurements. This comparison shows that, on average, ranges are reduced by 35% (for example where a simple calibrated date might span 300 years, the posterior density estimate spans only 200 years). Looking at this from the point of view of a Civil Service accountant, one could say that the value-for-money of the English Heritage radiocarbon budget has been increased by around 25%. More importantly radiocarbon dating, integrated with other forms of chronological evidence, now has the potential to tackle a much wider range of archaeological problems, which require precision previously outside the scope of the method.

2.2 Pragmatists and Archaeologists

Archaeologists are fundamentally comfortable with the Bayesian approach to interpreting archaeological data. The past is a foreign country where they do things differently, and we are used to looking in through a frosted window.

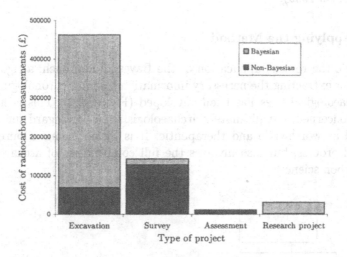

Fig. 2.1. Bar chart showing the cost of radiocarbon dates commissioned and analysed by English Heritage between 1994 and 2000, by project type and analytical approach.

The production of alternative models, none of them definitive, is simply one means of creating multiple pasts and is entirely congruent with a post-modern perspective (Shanks and Tilley 1987). The analysis of data using our 'prior beliefs' about the problem under consideration is simply an explicit, processual approach to contextual data (Clarke 1979; Hodder 1986). Archaeologists are used to fragments of data, and a plethora of views on how those data should be interpreted.

In addition to providing an approach which is theoretically attractive to many archaeologists, mathematical modelling also descended like manna from heaven. It provided an explicit methodology to do something which archaeologists have wanted to do for many years, but have not had the tools to accomplish. A random browse through any radiocarbon datelist of the 1970s or 1980s is enough to establish that archaeologists frequently have independent evidence about the chronology of their sites, but have been unable to use this quantitatively to affect their radiocarbon chronologies (see, for example, Jordan et al. 1994, 10).

However much archaeologists are attracted to the Bayesian concept, the practice is a different kettle of fish. "Statistics seems to be one of those subjects that cause instant mental paralysis in many otherwise competent archaeologists" (Fletcher and Lock 1991, viii). It is not so much the equations that are the problem, as the idea of equations. This problem has to be addressed because the financial and theoretical advantages of the methodology are so great (Orton 1999).

2.2.1 Applying the Method

To enable the routine application of the Bayesian approach, a rigorous procedure for extracting the necessary information to build chronological models for archaeological sites has been developed (Figure 2.2). This is hard work for all concerned, but ultimately archaeologists tend to regard the discipline involved as worthwhile and therapeutic. It is by no means a purely mathematical process, but also involves the full complexities of archaeology and radiocarbon science.

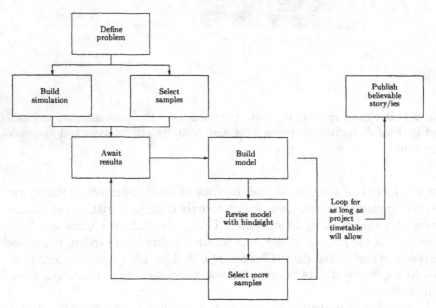

Fig. 2.2. Flow diagram showing stages in routine chronology building.

This process may be illustrated by an example – the area of the cemetery at Wharram Percy church occupied subsequently by the glebe. As part of a wider research study in the parish (Beresford and Hurst 1990), a major programme of excavations was carried out in and around the church of St Martin from 1950 to 1990. The excavations within the church have been fully published (Bell et al. 1987). The dating of burials in the glebe was undertaken as part of the wider post-excavation programme for the publication of the cemetery (Harding 1996), and we are grateful for permission to report these results in advance of the final site publication.

2.2.2 Formulating the Questions

The first stage of the chronology building process is to define what chronological questions are of interest to the project. This is a matter for the archaeologist. On the whole, this stage raises relatively few difficulties. The most common problem is providing a meaningful definition of 'contemporaneity'. Depending on context, "Is A contemporary with B?" may mean "Is it within the same decade?" or "Are both late Neolithic?" (in England a period covering most of the third millennium BC). Determining the resolution of chronology required if dating is to be archaeologically useful is critical to this stage of the process – there is no point wasting resources on questions which are unanswerable given the techniques, material, and resources available.

In the glebe at Wharram Percy the questions of interest were:

• when burial extended into this northern part of the cemetery
• how long this was after the start of burial on the site
• for how long burial in the area continued.

The date when burial ceased in this part of the site was of less interest, as the area was sealed by a boundary wall, thought to be related to the construction of a new vicarage in AD 1328.

The next parts of the process occur simultaneously, often forming a feedback loop. Firstly a simulation of the problem is built, based on the 'prior information' available about the dating of the samples, the archaeologist's estimated date for the site, the available samples, and the type of measurements which may be obtained from them.

The first stage in the simulation is to find the expected radiocarbon concentration of the events whose calendar ages have been defined. The expected dating uncertainty is then used to take a sample from a normal distribution about the true radiocarbon concentration. This 'simulated' radiocarbon measurement is then calibrated and analysed in the normal way.

Frequently, several simulations are built and compared. This may be necessary to compare different interpretations of the stratigraphy, or to determine whether high-precision measurements are justified, or to compare the effects of different parts of the calibration curve in cases where the actual date of the samples can be estimated only imprecisely.

2.2.3 Model Construction and Simulation of the Dating

A simulation of the dating of the burials from the glebe at Wharram Percy is shown in Figure 2.3. This is based on the preliminary evaluation of the chronological relationships between the events of interest for the site, an estimate that burial in this area continued from AD 900 to AD 1328, dating of 22 articulated skeletons (from a total of 118 burials with bone surviving), and errors likely to be obtained by conventional radiometric dating.

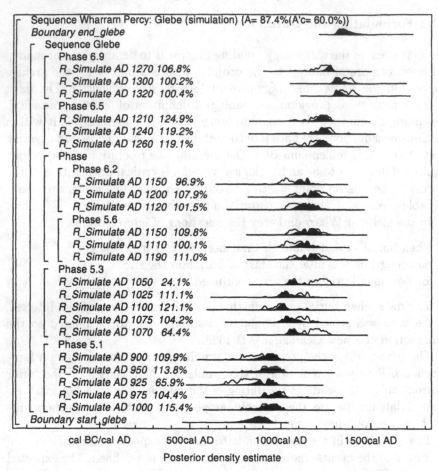

Fig. 2.3. Probability distributions of simulated radiocarbon dates from burials in the glebe. The expected radiocarbon concentration of a sample which actually dates to the calendar age defined (e.g. AD 1000) is calculated, and the expected error term for conventional radiocarbon measurements (±50 BP) is used. Each distribution represents the relative probability that an event occurred at a particular time. For each simulated radiocarbon date, two distributions have been plotted: one in outline which is the result of simple radiocarbon calibration, and a solid one based on the chronological model used. The other distributions correspond to aspects of the model. For example, the distribution '*start_glebe*' is the estimated date for the first burial in this area of the churchyard. The large square brackets down the left-hand side of the figure, along with the *OxCal* keywords, define the overall model exactly.

In this case the model suggests that we should be able to estimate the date when burial started in the area to within two hundred years. This is not really precise enough to answer our archæological questions, and so further models were built to determine whether high-precision measurements (Figure 2.4) or more measurements at conventional precision (Figure 2.5) would be more effective at refining this estimate. In fact both models produce estimates which span just over a century and both approaches were physically possible, so in this case the decision to obtain conventional radiocarbon dates was based on practical factors (i.e. cost and timescale).

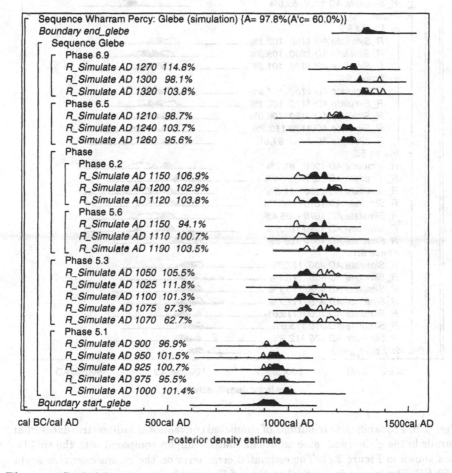

Fig. 2.4. Probability distributions of simulated high-precision dates from burials in the glebe. The format is identical to that for Figure 2.3, although the expected error term on the measurements is now ±18 BP.

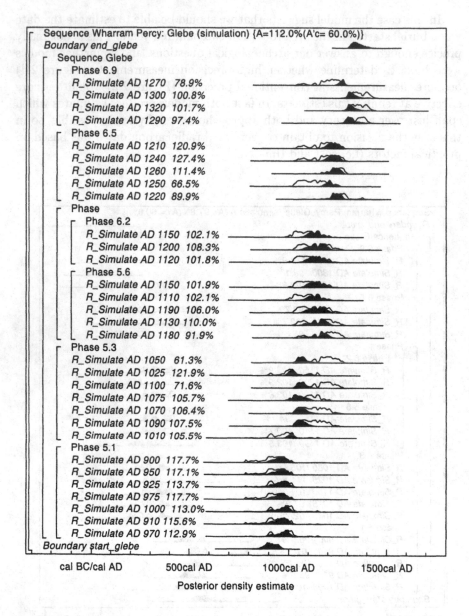

Fig. 2.5. Probability distributions of simulated conventional radiometric dates from burials in the glebe (with nine additional dated samples compared with the simulation shown in Figure 2.3). The estimated error term on the measurements is again ±50 BP. The format is identical to that of Figure 2.3.

2.2.4 Sample Selection

This stage of the process is generally more difficult.

Frequently the 'prior information' incorporated into the model is very informative, particularly as it is usually based on site stratigraphy rather than more nebulous concepts such as archaeological 'phasing'. There are occasions, however, when such information is limited and the results are much more sensitive to (supposedly) uninformative mathematical assumptions.

A second difficulty is archaeological. The relative dating provided by stratigraphy relates to the deposition of contexts. Radiocarbon measurements relate to samples. The absolute age of samples is not necessarily the same as the absolute age of the deposition of the contexts from which they derive. However, if the stratigraphic sequence is to be used as prior information for the interpretation of the date of the contexts and the samples within them, then these dates must be equivalent. To ensure this, a rigorous assessment of the taphonomic process (i.e. the means by which a sample enters an archaeological deposit or context) becomes critical. Every sample is residual until proven otherwise. The contextual integrity of radiocarbon samples has been a serious concern in sample selection for many years (Bowman 1990; van Strydonck et al. 1999), but this becomes even more critical when the contextual information is used to modify the chronologies produced.

Radiocarbon science also comes into play at this point. Not only must the sample have been fresh when it got into its context, but it must have ceased exchanging carbon with the atmosphere at that time. Technical factors such as collagen turnover, old-wood offsets, and components of marine carbon, also become significant (see, for example, Bowman 1990 for an introduction to these issues).

All these multi-disciplinary complexities have to be considered for every sample. But that is only the beginning. Archaeologists are used to selecting radiocarbon samples. They are used to receiving radiocarbon dates. With this methodology the interpretation of each sample depends on all of the others and we can also estimate things (such as the start of an archaeological phase) which are not directly dated at all. This is the concept which archaeologists find hardest to grasp. The concept of a 'radiocarbon date' is deeply ingrained, and the shift from dating samples to estimating chronology takes some getting used to.

Emerging from the mangle that is sample selection and the creation of a sampling strategy, either we have decided that radiocarbon dating cannot resolve our archaeological questions or we wait whilst the radiocarbon content of our samples is measured.

2.2.5 Analysis of the Results and Generation of a Working Model

On receipt of our measurements, it should be simply a matter of fitting the measured results into the model in place of the simulations, and then performing the analysis. Of course, it is never that simple! The results of the conventional radiometric measurements, simulated in Figure 2.3, are shown in Figure 2.6. One can see from the plots that the posterior distributions are not in good agreement with the calibrated radiocarbon dates themselves. This can be quantified for each event by calculating an overlap integral between the likelihood distribution (from the calibration itself) and the posterior distribution. The product of these overlaps gives a measure of the agreement between the model and the data. In *OxCal* these measures are used to calculate an agreement index (A) for each event and an overall agreement index for the model as a whole $(A_{overall})$. The indices are calculated so that values less than 60% generally indicate a high likelihood $(> 95\%)$ that there is a problem with the model (details of these indices are given in Bronk Ramsey 1995). This model has poor overall agreement index $(A_{overall} = 8.3\%)$, suggesting that the radiocarbon dates and the stratigraphic information incorporated in the model are contradictory.

In this specific case, all the samples were from articulated burials, so there is no question that the samples are of the same date as the contexts dated. This example is unusual, however, as it is this taphonomic relationship between sample and context which is usually the most fragile link in the chain of sample selection. Most often it is the interpretation of the taphonomy of the dated samples which needs to be re-examined, although more rarely the record of the stratigraphy or the radiocarbon measurements may be at fault. In the glebe, it was the preliminary phasing which came under scrutiny – in particular the allocation of disturbed burials to the latest possible archaeological phase into which they could fall. On this basis, a revised phasing scheme was proposed which produces the model shown in Figure 2.7. This model shows good agreement $(A_{overall} = 101.4\%)$, and provides an estimate for the start of burial in this area of the cemetery which is sufficiently precise to achieve the objectives of the dating programme. In this case, no further samples are needed and the project could proceed directly to publication. It is more usual, however, for the first set of samples to raise further questions and for the sample selection and model revision process to be repeated once or twice before a stable and consistent model is achieved.

In the real world, time, resources, or suitable samples may all run out before this happy state of affairs comes to pass. In a decade of trying, the authors have never created the perfect model for a site. Often we cannot even build the model which the project team thinks is closest to reality, because of the limitations of the 'module-based' software available. Ultimately this doesn't really matter, any model we build is just that – a *model* which represents a view of the past glimpsed by a certain set of people at a certain time through a portion of the frosted pane. What is important is that the models produced are likely to be more realistic than simply regarding the radiocarbon measurements in isolation. The team working on a post-excavation project are also likely to know the site data more intimately than anyone else is ever likely to, however accessible our archives, and to have access to resources for research and dating.

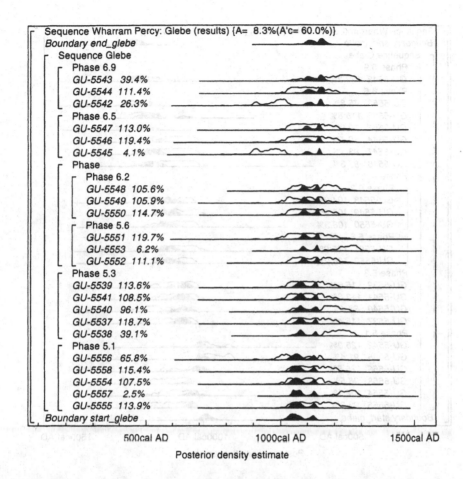

Fig. 2.6. Probability distributions of dates from burials in the glebe using the model based on the preliminary site phasing. The format is identical to that of Figure 2.3.

2.3 Fundamentalists and Physicists

One of the main problems in practice with the application of Bayesian statistics to archaeological chronology, is that absolute Bayesian rigour is almost impossible to achieve. Different aspects of this trouble different people: physicists brought up in a Classical statistical environment often don't like adding any arbitrary assumptions to an analysis; fundamentalist Bayesians might want to apply a probability to everything and let it all come out in the wash.

2.3.1 Limited Range of Models

In practice the ideal models for a site or culture are unlikely to be available in a pre-packaged form. For this reason there is a need to have models which

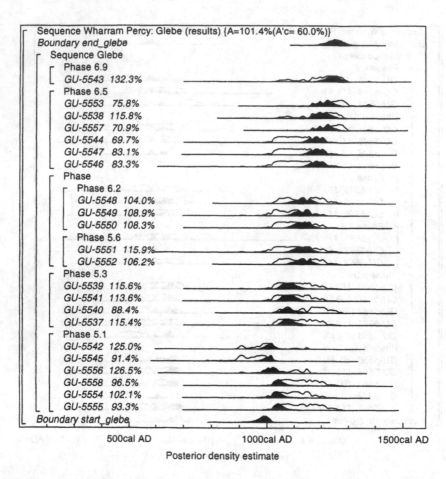

Fig. 2.7. Probability distributions of dates from burials in the glebe using the model based on the final site phasing. The format is identical to that of Figure 2.3.

are widely applicable and which are modular so that they can be built up to describe complex situations. This has been the approach with *OxCal* with the ability to nest phases (groups of events with no relative temporal relationship but sharing the same constraints) and sequences (events or groups of events which are constrained to be in a particular order), and add *termini ante quem*, *termini post quem* etc. (Bronk Ramsey 1998). This does allow a large range of sites to be analysed but inevitably you are working with what is available, not with what is ideal. The most obvious example of a limitation is the Poisson process for events between group boundaries (or uniform random distribution of events) which is assumed in most contexts. In practice you might often expect such distributions to be non-uniform (e.g. more activity at the start). This can be approximately modelled by using sub-groups but this is still only an approximation. In theory a program could be written for each site or culture which was able to reflect all of the quantified information more precisely, but this is not feasible in practice.

2.3.2 Uninformative Priors

The next problem, which is, in some senses, a manifestation of the previous one, is that the models themselves do impose constraints on the posterior distributions which may be unintentional. The classic example is the case of many age determinations on objects from a single phase. The more measurements that are made, the earlier will be the first and the later will be the last. If the dates are treated in a simplistic fashion assuming that they are totally independent and unrelated, you would conclude that the phase was longer the more measurements you made. Putting this in numerical terms: if the span of n events is s, the number of possible combinations of event dates scales as s^{n-2}. This is true whether or not the dates are supposed to be in a sequence. Thus, if the overall length of the group is not given a uniform prior (or some other appropriate weighting) the natural statistical weight of the events will bias (very strongly) to long phases (Steier and Rom 2000). Traditionally this problem has been fudged by the use of floruits (a form of inter-quartile range) and other ad-hoc evaluations of the data.

Although this particular problem can be properly treated using Bayesian analysis (Buck et al. 1992; Bronk Ramsey 2000, 2001), other less severe problems do remain. One example is that a large number of phases will tend to spread themselves roughly uniformly through time given no other dating information. This means that the duration of a whole chronological model will be biased to longer spans. This can also be addressed by applying a prior to the overall span (Bronk Ramsey 2001; Nicholls and Jones 2001; Sahu this volume). Another is that errors containing a systematic component (either through over-estimation of errors or, for example, with marine reservoir offsets) are usually not treated correctly, under-estimating spreads (Jones and Nicholls 2001).

One needs to be aware of these problems. In practice one of the best approaches we have found for tackling these difficulties is to look at the bare models with no likelihoods added (or at least with only overall constraining ones). This enables you to see the extent to which your posterior distributions are constrained by your chronological relationships and prior assumptions on spans within the model rather than by your likelihoods (Bronk Ramsey 2001). Another approach is to look at slightly different models and see how sensitive the conclusions are (Steier et al. 2001; Lu et al. 2001).

Perhaps ultimately this is non-Bayesian. If your model is right you should not mind whether the constraints on the posterior distributions come from the model or from the measurements. The point we are making is that models are almost never perfect and can have unintentional side-effects.

2.3.3 The Need for Clarity

Arising from the problems outlined above is a need to have a good intuitive feel for what is involved in the model that has been applied. We feel that such a model should be specified in a way which is both mathematically precise and as clear as possible to a non-specialist. For this reason we have tried to use a sequence and phase nomenclature recognizable to archaeologists. Because the models are not perfect, and because there is no 'right' model anyway, we might need to explore more than one model for a particular site. The implications of the different models should be clear to those reading the conclusions of the analysis.

One implication of this is undoubtedly controversial. If we draw up a model with a particular sequence, for example, it is fairly clear what this model implies. We are much less sure if anyone really has an instinct for what the implications of assigning probabilities to each sample being in sequence are – although the mathematical formulation of this certainly can be rigorous.

In practice this often fits well with the nature of archaeological evidence since 'prior information' which is incorporated into models is often sharply defined (e.g. Sample A is earlier than Sample B). The reason for this is that any quantification of the probability that this relative relationship is wrong is capricious. True stratigraphic relationships are definite: A is either earlier, or later, or precisely contemporary with B. Reality has no uncertainty. It is our knowledge of the relationship which has uncertainty, and this is fundamentally a qualitative rather than a quantitative variable. If Sample A is an articulated skeleton the legs of which have been truncated by the cut for Sample B, another articulated skeleton, then we are very confident of the relationship. If the grave cut for B simply overlaps that for A by 5 cm and doesn't disturb the bones of A, then we are less confident. Less confident, not 90% confident in the first case and 5% confident in the second. Attaching numbers converts an ordinal variable into a continuous one arbitrarily. This is at the heart of some of the initial reservations about the use of the Bayesian approach in archaeological reasoning (Reece 1994).

2.3.4 Hybrid Approach

One interpretation of this approach is that it is essentially a hybrid one. We are using Bayesian mathematics to combine probability distributions. However in building our models we are not doing so in the way a fundamentalist Bayesian might. Rather, we are taking the information we have from sites and cultures and then building that into a simplified prior model. This is the hypothesis on which the interpretations are based. As with Classical hypotheses, these models should be relatively simple without too many unquantifiable factors if any clarity is to be preserved. A good example of this is the analysis of the British Museum Bronze Age material (Needham et al. 1998); here the data have been modelled using one model where it is assumed that the types of bronze work belonged to independent and possibly overlapping phases, and another where there are sequential phases. Archaeologically speaking, there are arguments for and against both of these models, but the authors feel that the comparison of the results of both analyses is more useful than applying what would have to be an arbitrary probability of overlap and using only one grand model.

We also have to admit to some instances of hypothesis testing (for example Holme-next-the-Sea; Bayliss et al. 1999): using the overlap between likelihood distributions and posterior distributions (an approximation to pseudo-Bayes-factors) to identify unlikely hypotheses, or unlikely aspects of a hypothesis. Interestingly, in cases where we have tested this approach, it identifies the same outliers as either the Classical chi-squared test (in simple combinations of normal distributions) or other Bayesian outlier tests, indicating that although the mathematical formulations are different, all of these approaches do give similar results in practice. For example, the date of the Holme-next-the-Sea timber circle has been subsequently – and independently – confirmed by dendrochronology (Brennand and Taylor forthcoming).

Ultimately the reason why we feel that this approach is useful in this particular context is because of the very complex and difficult nature of the evidence. If probabilities could be accurately assigned to all aspects of a model, and perfect models that really fitted the prior knowledge could be constructed, a pure Bayesian approach would make sense. In these cases we almost never have all the information we would need to build such a model anyway. In practice the models are themselves only an imperfect simplification of the true situation and so we feel that keeping them simple, and treating them as hypothetical assumptions in our interpretation better reflects the reality of our intellectual approach.

2.4 Conclusions

The last decade has been a period of rapid advance in the routine application of Bayesian statistics to chronology. Bridging differences in theoretical approach and understanding between archaeologists, natural scientists, and mathematicians can require considerable effort from all concerned. Our approach towards the practical implementation of Bayesian chronological modelling on a routine basis is both eclectic and pragmatic (Steel 2001). We feel that the results undoubtedly justify the means, and that with further methodological and scientific developments, the technique should be even more widely applicable.

References

Bayliss, A., Groves, C., McCormac, F. G., Baillie, M., Brown, D. and Brennand, M. (1999). Precise dating of the Norfolk timber circle. *Nature*, **402**,

479.

Bell, R. D., Beresford, M. et al. (1987). *Wharram Percy: the church of St Martin*. Monograph Series, **11**. Society for Medieval Archaeology, London.

Beresford, M. and Hurst, J. (1990). *English Heritage book of Wharram Percy deserted medieval village*. Batsford, London.

Bowman, S. (1990). *Radiocarbon dating*. British Museum Press, London.

Brennand, M. and Taylor, M. (forthcoming). The survey and excavation of a Bronze Age timber circle at Holme-next-the-Sea, Norfolk, 1998–9. *Proceedings of the Prehistoric Society*.

Bronk Ramsey, C. (1995). Radiocarbon calibration and analysis of stratigraphy: the OxCal program. *Radiocarbon*, **37**, 425–430.

Bronk Ramsey, C. (1998). Probability and dating. *Radiocarbon*, **40**, 461–474.

Bronk Ramsey, C. (2000). Comment on 'The use of Bayesian statistics for ^{14}C dates of chronologically ordered samples: a critical analysis'. *Radiocarbon*, **42**, 199–202.

Bronk Ramsey, C. (2001). Development of the radiocarbon calibration program OxCal. *Radiocarbon*, **43**, 355–363.

Buck, C. E., Kenworthy, J. B., Litton, C. D. and Smith, A. F. M. (1991). Combining archaeological and radiocarbon information: a Bayesian approach to calibration. *Antiquity*, **65**, 808–821.

Buck, C. E., Litton, C. D. and Smith, A. F. M. (1992). Calibration of radiocarbon results pertaining to related archaeological events. *Journal of Archaeological Science*, **19**, 497–512.

Clarke, D. L. (1979). *Analytical archaeology*. Academic Press, London.

Fletcher, M. and Lock, G. (1991). *Digging numbers: elementary statistics for archaeologists*. Oxford University Committee for Archaeology, Monograph **33**, Oxford.

Harding, C. (1996). Wharram Percy (North Yorkshire): topography and development of the churchyard. In H. Galinié and E. Zadora-Rio (eds.), *Archéologie du cimetière chrétien: actes du 2e colloque ARCHEA*. FERACF/La Simarre, Tours, 183–191.

Hedges, R. E. M., Housley, R. A., Bronk Ramsey, C. and van Klinken, G. J. (1992). Radiocarbon dates from the Oxford AMS system: Archaeometry datelist 15. *Archaeometry*, **34**, 337–357.

Hodder, I. (1986). *Reading the past: current approaches to interpretation in archaeology*. Cambridge University Press, Cambridge.

Jones, M. and Nicholls, G. (2001). Reservoir off-set models for radiocarbon calibration. *Radiocarbon*, **43**, 119–124.

Jordan, D., Haddon-Reece, D. and Bayliss, A. (1994). *Radiocarbon dates from samples funded by English Heritage and dated before 1981*. English Heritage, London.

Lu, X., Guo, Z., Ma, H., Yuan, S. and Wu, X. (2001). Data analysis and calibration of radiocarbon dating results from the cemetery of the Marquises of Jin. *Radiocarbon*, **43**, 55–62.

Needham, S., Bronk Ramsey, C., Coombs, D., Cartwright, C. and Pettitt, P. B.
(1998). An independent chronology for British Bronze Age metalwork: the
results of the Oxford Radiocarbon Accelerator Programme. *Archaeological
Journal*, **154**, 55–107.

Nicholls, G. and Jones, M. (2001). Radiocarbon dating with temporal order
constraints. *Applied Statistics*, **50**, 503–521.

Orton, C. R. (1999). Plus ça change – perceptions of archaeological statistics. In L. Dingwall, S. Exon, V. Gaffney, S. Laflin and M. van Leusen
(eds.), *Archaeology in the age of the Internet: proceedings of the conference
on computer applications and quantitative methods in archaeology, 1997*.
British Archaeological Reports, International Series, **S750**, 25–34.

Reece, R. (1994). Are Bayesian statistics useful to archaeological reasoning?
Antiquity, **68**, 848–850.

Rom, W., Golser, R., Kutschera, W., Priller, A., Steier, P. and Wild, E. M.
(1999). AMS ^{14}C dating of equipment from the iceman and of spruce logs
from the prehistoric salt mines of Hallstatt. *Radiocarbon*, **41**, 183–197.

Sahu, S. K. (this volume). Applications of formal model choice to archaeological chronology building. In C. E. Buck and A. R. Millard (eds.), *Tools
for constructing chronologies: crossing disciplinary boundaries*, Springer-
Verlag, London.

Shanks, M. and Tilley, C. (1987). *Social theory in archaeology*. Polity Press,
Oxford.

Steel, D. (2001). Bayesian statistics in radiocarbon calibration. *Philosophy of
Science*, **68**, S153–S164.

Steier, P. and Rom, W. (2000). The use of Bayesian statistics for ^{14}C dates
of chronologically ordered samples: a critical analysis. *Radiocarbon*, **42**,
183–198.

Steier, P., Rom, W. and Puchegger, S. (2001). New methods and critical
aspects in Bayesian mathematics for ^{14}C calibration. *Radiocarbon*, **43**,
373–380.

van Strydonck, M., Nelson, D. E., Crombe, P., Bronk Ramsey, C., Scott,
E. M., van der Plicht, J. and Hedges, R. E. M. (1999). What's in a ^{14}C
date. In J. Evin, C. Oberlin, J. P. Daugas and J. F. Salles (eds.), *^{14}C et
archéologie: 3ème congrès international, Lyon, 6–10 avril 1998*. Mémoires
de la Société Préhistorique Française Tome XXVI et Supplément 1999 de
la Revue d'Archéometrie, 433–440.

3

Bayesian Inference of Calibration Curves: Application to Archaeomagnetism

Philippe Lanos

Summary. This chapter focuses on recently developed models for the analysis and interpretation of archaeomagnetic dating evidence. Archaeomagnetic data from archaeological structures such as hearths, kilns or sets of bricks and tiles, exhibit considerable experimental errors and are typically also associated with date estimates from other sources such as stratigraphic sequences, historical records or chronometric methods. This chapter summarizes the technical aspects of recent Bayesian statistical modelling work, describing a hierarchical model for the archaeomagnetic data and its uncertainties and combining this with models of the other dating evidence, based on those described by Buck (Chapter 1), to create a calibration curve for future archaeomagnetic dating work in a locality. The proposed model and inference methods are illustrated by the construction of a calibration curve using recently published archaeomagnetic data from Lübeck, Germany. With this new posterior estimate of the curve available, it is then possible to use the Bayesian statistical framework to estimate the calendar dates of undated archaeological features.

3.1 Introduction

In this chapter, I summarize recent work aimed at devising a formal Bayesian framework for the construction of chronologies based on archaeomagnetic dating. Archaeomagnetic dating utilizes the property of some materials to record or 'fossilize' information about the Earth's magnetic field (EMF). This 'fossilized' magnetic information is known as remanent magnetism and is acquired, for example, by heating materials at high temperatures. Typically, archaeomagnetic data relate to the direction (inclination, I, and declination, D) and intensity (denoted F for magnetic field) and so a complete archaeomagnetic record is three-dimensional. At any given time, these three components of the EMF vary according to geographical location (latitude and longitude). Moreover, at a given place, these three components vary through time. The archaeomagnetic dating method, as applied to baked clay, originated in France in the 1930s, thanks to the early work by Thellier (Thellier 1938, 1981; Thellier and Thellier 1959), but it is really since the 1960s that the method has reliably been applied to dating archaeological kilns and ovens.

It is now well established that, if we study a sufficient number of archaeological structures in a given area that relate to a given archaeological period, then it is possible to build local secular variation curves for each of the three EMF variables. If these curves are known with sufficient accuracy, it is then possible to date other structures in the same geographical region on which magnetization has been measured. Thus, for some countries, the archaeomagnetic method allows dating of samples up to several millennia old (see, for instance, Tarling 1983; Sternberg 1989; Bucur 1994; Kovacheva and Toshkov 1994; Marton 1996; Kovacheva 1997; Batt 1997; Kovacheva et al. 1998; Lanos et al. 1999; Chauvin et al. 2000; Gallet et al. 2002).

The theory presented here is illustrated throughout the chapter by an outstanding data set presented in Section 3.2. Section 3.3 consists of a description and statistical analysis of the archaeomagnetic data which leads to derivation of a Bayesian hierarchical model. In Section 3.4 I will propose a general method for estimating calibration curves using the Bayesian approach. The aim is to estimate the geomagnetic parameter(s) studied (inclination and/or declination and/or intensity) and also to define, as reliably as possible, highest posterior density (HPD) intervals on these estimates. With the new posterior estimate of the curve available, in Section 3.5, I show how the Bayesian statistical framework can be used to estimate the calendar dates of undated archaeological features (such as kilns).

3.2 A Tutorial Example from Lübeck (Germany)

Consider data relating to a sequence of floors from a medieval bread oven investigated by Schnepp et al. (2003). In 1987–1988, archaeologists conducted excavations in a bakery at Mühlenstrasse Plot 65 in Lübeck, Germany, in advance of a building renovation project on the site (Müller 1992). The town of Lübeck, former head of the Hanseatic League, has existed since the early 12th century, and the bakery was situated in one of its oldest sections, on the Kings' Highway (the main road leading south). The archaeological excavations uncovered seven ovens in Plot 65, each built upon its predecessor. All the eastern halves were destroyed during the investigation, but the western halves were preserved, still allowing access to individual floors (Figure 3.1).

Inside the domes of the ovens, several consecutive floors of superficially baked loam (Schnepp and Pucher 1998) could be attributed to each building phase, except for the oldest phase II, where the floors had been destroyed while constructing the next oven. There are 25 floors (3 to 15 cm thick, phases III to VI, see Figure 3.1) still extant, adding up to a height of roughly 1.1 m. Since the fires to heat the oven burned on these floors, their smooth finish would have become rough, making the bread stick, so new floors were added regularly.

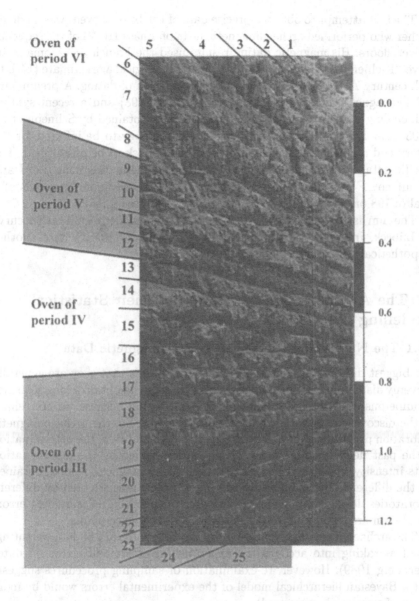

Fig. 3.1. Photograph of the Lübeck oven-floor sequence taken during the excavation in 1987 (modified after Schnepp et al. 2003). A sequence of 25 oven floors is seen (phases III to VI). Each oven floor can be easily distinguished because of the smooth and dense yellow-grayish surface, while the bottom part is formed of crumbly, reddish baked clay. The oven floors were numbered as shown, starting at the top (period VI).

The first attempt to obtain a precise date of the bakers' ovens was made by Pucher who performed archaeomagnetic tests on phase III–VI of the sequence of oven floors. His magnetic dating results used the French archaeomagnetic curve (Thellier 1981) and agreed with the archaeological age estimate (14th to 17th century AD) but did not increase the precision of dating. A preliminary archaeomagnetic analysis (Schnepp and Pucher 1998) and a recent synthesis including new archaeomagnetic dating results obtained by Schnepp et al. (2003) now allow a more precise reference chronology to be inferred for the bakery and a geomagnetic reference curve for Lübeck to be established. The specific data available take the form of palaeomagnetic directions (i.e. I and D, but not F) that have been determined for 24 oven floors by means of a total of 198 oriented samples (Schnepp et al. 2003; see Table 3.1).

The aim is to construct a directional archaeomagnetic calibration curve for Lübeck, improve the dating of the floors and be able to date another (hypothetical) structure of unknown age.

3.3 The Archaeomagnetic Data and their Statistical Modelling

3.3.1 The Nature and Quality of Archaeomagnetic Data

The biggest disadvantage of archaeomagnetic data is that they are generally unevenly distributed in time and space. It is difficult to obtain a long sequence of numerous and well-dated data for a given region because we are reliant on the discovery of suitable archaeological remains. So, the archaeomagnetic calibration process is a long and exacting task. In addition, the determination of the past field direction is not always accompanied by the determination of its intensity and vice versa. Finally, the archaeomagnetic results, obtained via the different sampling and measurement procedures adopted by different laboratories, lead to various precisions which depend on measurement errors but also on errors relating to the dates used as reference points.

The archaeomagnetic literature proposes different types of data weighting, aimed at taking into account the experimental errors and errors on dates (Sternberg 1989). However, re-examination of sampling procedures suggests that a Bayesian hierarchical model of the experimental errors would be more robust (Lanos et al. submitted).

Generally, we estimate the ancient magnetic field at a given place and time from a number of contemporaneous archaeological baked clay structures (kilns, ovens, etc. – in the case of Lübeck, these correspond to the floors). One gathers from each structure a set of independently oriented samples with geographical coordinates. The samples are generally cut into one or more cubic or cylindrical specimens that can be measured in spinner magnetometers. Thus, I define hierarchical levels called 'specimens', 'samples' and 'structures' (Figure 3.2, box labelled 'likelihood'), each relating to different sets of error sources.

Table 3.1. Mean remanent magnetization directions of the Lübeck oven-floor sequence situated at 53.87°N and 10.81°E in Germany (after Schnepp and Pucher 1998; Schnepp et al. 2003). Column legend: *Floor*: floor number (here equivalent to 'structure' level in the hierarchical sampling model, that is $m_i = 1$), I_{ij}: mean inclination of the floor, D_{ij}: mean declination of the floor, K_{ij}: concentration factor, r_{ij}: number of specimens, t_{ij1}: latest calendar date (prior TPQ), t_{ij2}: earliest calendar date (prior TAQ), *Prior dating*: fg-TL: fine grain thermoluminescence, cg-TL: coarse-grain TL, AMS C14: radiocarbon date with accelerator mass spectrometry, historical: historical date from texts, stratigraphic: stratigraphic constraint. All the floors are ordered in sequence from floor 25 (oldest) to floor 1 (youngest).

Floor	$I_{ij} = I_i$	$D_{ij} = D_i$	K_{ij}	r_{ij}	$t_{ij1} = t_{i1}$ (prior TPQ)	$t_{ij2} = t_{i2}$ (prior TAQ)	Source of prior dating evidence
1	73.90	344.00	595	14	1791	1804	historical
2	73.90	355.70	1712	9	1628	1804	stratigraphic
3	76.30	352.10	2264	8	1628	1804	stratigraphic
4	76.90	355.50	1570	9	1628	1804	stratigraphic
5	76.10	354.50	580	8	1628	1804	stratigraphic
6	75.10	358.10	1146	7	1628	1804	stratigraphic
7	76.40	356.90	1210	9	1628	1804	stratigraphic
8	72.60	13.70	363	9	1628	1804	stratigraphic
9	71.70	12.20	332	6	1628	1804	stratigraphic
10	70.80	16.00	329	8	1628	1740	fg-TL
12	70.10	15.00	640	7	1517	1709	fg-TL
13	71.10	12.80	1498	4	1502	1709	stratigraphic
14	69.30	14.60	600	9	1502	1654	cg-TL
15	68.30	15.10	559	9	1449	1654	stratigraphic
16	69.60	8.10	239	8	1449	1608	fg-TL
17	68.20	3.60	223	8	1484	1596	fg-TL
18	67.00	10.60	1306	8	1448	1608	cg-TL
19	66.60	6.90	894	9	1428	1549	fg-TL
20	65.30	3.90	356	10	1301	1549	stratigraphic
21	63.50	7.70	518	7	1301	1549	fg-TL
22	64.50	3.40	162	8	1324	1476	fg-TL
23	64.70	4.90	1207	9	1304	1500	fg-TL
24	63.60	9.50	445	8	1284	1500	stratigraphic
25	65.90	14.20	146	8	1284	1378	AMS-C14, fg-TL and historical

3.3.2 A Hierarchical Modelling Approach to Archaeomagnetic Data

Here, I propose a Bayesian hierarchical model for archaeomagnetic data. In the case of Lübeck, specimens are assimilated to samples. This leads to simplifications within the general model presented in Lanos et al. (submitted).

For a particular 'sample', characterized by the indices ijk (i for time, j for structure, k for sample), the archaeomagnetic laboratory determines a direction (I_{ijk}, D_{ijk}). This direction includes the measurement and different magnetic corrections. A 'sample' is a block of burnt clay gathered from a structure, or in the case of 'displaced' materials, a fragment of brick or pottery (Lanos et al. 1999). The term 'sample' is not taken here in its statistical sense as a set of observations. It describes a piece of clay gathered in such a way as to record the precise conditions of its orientation.

The 'structure' level represents a furnace, oven, burnt earth or wall, layer, floor, or a homogeneous set (same production) of displaced material (tiles, bricks, pottery), i.e. "volume of material that can be considered to have been magnetized at the same time" (Tarling 1983). A structure ij possesses an unknown mean direction (I_{ij}, D_{ij}) corresponding to a time t_i. The common characteristics are the geomagnetic field elements determined by the acquired magnetization at a given time t_i expressed in the calendar (or so-called solar or sidereal) date system. Several structures ($j = 1, \ldots, m_i$) at time t_i define a 'tied' design point at time knot t_i (Green and Silverman 1994, p. 43). A weight P_{ij}, such that $\sum_{j=1}^{m_i} P_{ij} = m_i$, can be attributed to each structure, that is 'floor' in the case of Lübeck, possibly depending on the measurement process. For example, certain geomagnetic (inclination, declination or intensity) determination processes can be considered more reliable than others (Chauvin et al. 2000). If unspecified, this parameter P_{ij} is set equal to one, and this will be the case in the rest of this chapter.

We finally introduce a magnetic 'field' level as a function of time $g(t_i) = g_i$, which is the ultimate direction to be estimated.

3.3.3 Details of the Hierarchical Modelling

At each level in the hierarchy, I model the archaeomagnetic directional data (i.e. the inclination and declination components) as Fisher distributed (Fisher 1953). A justification of the use of Fisher's statistics for archaeomagnetic data can be found in Love and Constable (2003), who represent the variance, σ^2, of palaeomagnetic vectors, at a particular site and of a particular polarity, by a probability density function in a cartesian space of three orthogonal magnetic field components consisting of a non-zero mean symmetrical Gaussian function. In the geophysically relevant limit of small relative variance σ/g_F (g_F being the total field intensity), they demonstrate that the directional distribution of the off-axis angle approaches a Fisher distribution and that the intensity distribution (of Rayleigh–Rician type) approaches a normal distribution. My analysis in this chapter adopts these last distributions which adapt well to archaeomagnetic data.

The relationship between the measured directions (I_{ijk}, D_{ijk}) at sample level and the hierarchical empirical direction $(\bar{I}_i, \bar{D}_i)$ at field level is established through an integration with respect to the unknown mean direction (I_{ij}, D_{ij}) (Lanos et al. submitted). This leads to the Bayesian hierarchical (empirical, or observed) mean direction

$$\begin{aligned} \bar{I}_i &= \arccos(\bar{z}_i) & 0° \le \bar{I}_i \le 180° \\ \bar{D}_i &= \arctan(\bar{y}_i/\bar{x}_i) & 0° \le \bar{D}_i \le 360° \end{aligned} \quad (3.1)$$

where

$$\bar{x}_i = \sum_{i=1}^{m_i} P_{ij}\bar{x}_{ij} \Big/ \sum_j P_{ij} \qquad \bar{x}_{ij} = \tfrac{1}{r_{ij}} \sum_{k=1}^{r_{ij}} x_{ijk} \qquad x_{ijk} = \cos I_{ijk} \cos D_{ijk}$$

$$\bar{y}_i = \sum_{i=1}^{m_i} P_{ij}\bar{y}_{ij} \Big/ \sum_j P_{ij} \qquad \bar{y}_{ij} = \tfrac{1}{r_{ij}} \sum_{k=1}^{r_{ij}} y_{ijk} \qquad y_{ijk} = \cos I_{ijk} \sin D_{ijk}$$

$$\bar{z}_i = \sum_{i=1}^{m_i} P_{ij}\bar{z}_{ij} \Big/ \sum_j P_{ij} \qquad \bar{z}_{ij} = \tfrac{1}{r_{ij}} \sum_{k=1}^{r_{ij}} z_{ijk} \qquad z_{ijk} = \sin I_{ijk},$$

the number of samples is r_{ij}, the number of structures is m_i, and bars on variate symbols indicate 'empirical average'.

The distribution of hierarchical empirical mean direction $(\bar{I}_i, \bar{D}_i)$ at field level, is a hierarchical Fisher statistic around the unknown direction of the magnetic field (g_{I_i}, g_{D_i})

$$p(\bar{I}_i, \bar{D}_i) = \frac{m_i K_{B_i}}{2\pi \sinh(m_i K_{B_i})} \times \cos \bar{I}_i$$
$$\times \exp\left[m_i K_{B_i}\left(\cos \bar{I}_i \cos g_{I_i} \cos\left(\bar{D}_i - g_{D_i}\right) + \sin \bar{I}_i \sin g_{I_i}\right)\right] \quad (3.2)$$

where K_{B_i} is the 'concentration factor', equal to the inverse of variance, which characterizes the dispersion of the empirical mean direction around the true (unknown) mean direction (Fisher 1953).

Using the Fisher statistics at each level according to a hierarchical Bayesian approach (Lanos et al. submitted), the so-called 'Bayesian hierarchical' concentration factor, K_{B_i}, appears as the inverse of a harmonic mean of variances $1/K_i$ and $1/K_{ij}$ which are defined at field and structure levels respectively

$$K_{B_i} = \frac{1}{m_i} \sum_{j=1}^{m_i} P_{ij} \left(\frac{1}{K_i} + \frac{P_{ij}}{r_{ij} K_{ij}} \right)^{-1}. \quad (3.3)$$

If the numbers of samples r_{ij} tends to infinity, then $K_{B_i} \to K_i$, which characterizes the dispersion between structure directions at time t_i.

The concentration factor K_{ij} (see Table 3.1) is estimated from the unbiased (hence the symbol *) empirical concentration factor thus

$$K_{ij}^* = \left(\frac{r_{ij}-1}{r_{ij}}\right)\frac{1}{1-\bar{R}_{ij}}, \tag{3.4}$$

where the resultant mean length, $\bar{R}_{ij}$, is given by $\bar{R}_{ij} = \sqrt{\bar{x}_{ij}^2 + \bar{y}_{ij}^2 + \bar{z}_{ij}^2}$.

The error cone at the 95% confidence level around the empirical mean at the structure level is determined (in radians) by

$$a_{95ij} = t_{\beta F}(r_{ij}) \left/ \sqrt{r_{ij}K_{ij}}, \right. \tag{3.5}$$

where $t_{\beta F}(x) = \sqrt{2(x-1)(\beta^{-1/(x-1)} - 1)}$ for $x \geq 2$ (Fisher 1953). When x tends to infinity $t_{\beta F} \cong 2.45$ (in fact, this value is nearly achieved for $x > 10$).

3.3.4 Lübeck Data

In the Lübeck example, one floor corresponds to one structure and to one time $t_{ij} = t_{i\bullet} = t_i$ (where index $\bullet$ means sum on j), that is $m_i = 1$. Consequently, weighting parameter P_{ij} is equal to one and, from Equations 3.1, the mean direction observed at field level is directly given by the mean direction observed at structure level: $(\bar{I}_i, \bar{D}_i) = (\bar{I}_{ij}, \bar{D}_{ij})$. The unknown inter-structure variance $1/K_i$ will be supposed constant over the entire calibration curve, and thus equal to a global variance $\sigma_\alpha^2 = 1/K_\alpha$ which will be evaluated during curve estimation (see Equation 3.22 and associated text). Thus, in the case of Lübeck, the 'Bayesian' hierarchical concentration factor K_{B_i} in Equation 3.3 becomes

$$K_{B_i} = \left(\sigma_\alpha^2 + \frac{1}{r_{ij}K_{ij}}\right)^{-1}.$$

3.4 Bayesian Inference of Calibration Curves

3.4.1 Towards Geomagnetic Calibration Curves

The mean direction $(\bar{I}_i, \bar{D}_i)$ estimated at each time t_i is a point estimation of the magnetic field. If one wanted to take into account the physical connection between the geomagnetic elements through time, an additional hypothesis about the nature of $g(\cdot)$, the path on the unit sphere, should be introduced into the statistical approach.

The issue of the preliminary choice of the function $g(\cdot)$ as a 'penalized spline' has already been discussed (Jupp and Kent 1987; Wahba 1990;

Tsunakawa 1992; Green and Silverman 1994), but is very rarely implemented in archaeomagnetism, essentially because of the apparent mathematical complexity. However, I believe that this tool is one of the best ways for obtaining a good estimation of a calibration curve. The next section proposes a general method for estimating the calibration curve and details its implementation in the specific case of archaeomagnetism.

3.4.2 A General Bayesian Chronometric Equation

A statistical model assumes a mathematical relation between observed data and parameters which describe the studied physical (or other) phenomena as well as errors which can affect these phenomena. The aim is to estimate parameters of interest, that is those which can provide relevant answers to our queries, considering observed data and potential errors (Buck et al. 1996; Lanos 2001). Thus, one records the data and the goal is to find the corresponding model (the calibration curve, the calendar dates, ...). In geophysics this task is called the inverse problem (Tarantola 1987; Ulrych et al. 1992).

We propose here a general chronometric equation defined in the Bayesian framework, which can combine data from different methods, prior dates and parameters, and relative chronology (stratigraphy), in order to make posterior inference about calibration curves and calendar dates.

Having defined

- two (or more) vectors of n independent measurements,

$$Y = \left((\bar{I}_1, \bar{D}_1), \ldots, (\bar{I}_n, \bar{D}_n) \right)^T$$

in the case of observed archaeomagnetic directions, and

$$M = (M_1, \ldots, M_n)^T,$$

another set of chronometric measurements (exponent T means the transpose of a vector);
- the associated time vector $t = (t_1, \ldots, t_n)^T$;
- the reference curves, $g_Y = \left((g_{I_1}, g_{D_1}), \ldots, (g_{I_n}, g_{D_n}) \right)^T$ in the case of archaeomagnetic field directions, and $g_M = (g_{M_1}, \ldots, g_{M_n})^T$ corresponding to measurements M;
- the stratigraphic constraints S;
- the global variance σ_α^2 on the curve g_Y;

Bayes' theorem can be written thus:

'Likelihood' and 'prior' densities

$$p(Y, M, g_Y, g_M, \sigma_\alpha^2, S, t) = p(Y, M | g_Y, g_M, \sigma_\alpha^2, S, t)$$
$$\times p(g_Y, g_M | \sigma_\alpha^2, S, t)$$
$$\times p(\sigma_\alpha^2 | S, t) \times p(S | t) \times p(t) \qquad (3.6)$$

are equal to 'posterior' and 'marginal' densities

$$= p(g_Y | Y, M, \sigma_\alpha^2, S, t) \times p(g_M | Y, M, S, t)$$
$$\times p(\sigma_\alpha^2 | Y, M, S, t) \times p(t | Y, M, S)$$
$$\times p(S | Y, M) \times p(Y) \times p(M). \qquad (3.7)$$

In Equation 3.6, the 'likelihood' and 'prior' densities are

- $p(Y, M | g_Y, g_M, \sigma_\alpha^2, S, t)$: distribution of the observed data (likelihood)
  - Variable Y represents archaeomagnetic or palaeomagnetic directions obtained from hierarchical modelling outlined in Section 3.3 (vector data).
  - Variable M represents radiocarbon ages or any other measurement related to time (scalar data).
- $p(g_Y, g_M, \sigma_\alpha^2, S, t)$: prior information, decomposed into
  - $p(g_Y, g_M | \sigma_\alpha^2, S, t)$: prior on the nature of the calibration curves needed to relate Y and M to t.
  - $p(\sigma_\alpha^2 | S, t)$: prior on global variance. Index α indicates that the global variance is related to the smoothing parameter α (see Appendix A).
  - $p(S | t)$: stratigraphic constraints. S is a qualitative statistical variable describing the different order relations on time vector t. There are $n!$ different orders.
  - $p(t)$: prior date information. Uniform or more complicated distributions from historical and other sources.

If Y, g_Y and σ_α^2 are assumed independent of M and g_M, then

$$p(Y, M | g_Y, g_M, \sigma_\alpha^2, S, t) \times p(g_Y, g_M | \sigma_\alpha^2, S, t) =$$
$$p(Y | g_Y, \sigma_\alpha^2, S, t) \times p(M | g_M, S, t) \times p(g_Y | \sigma_\alpha^2, S, t) \times p(g_M | S, t).$$

In Equation 3.7, 'posterior' and 'marginal' densities are

- $p(g_Y | Y, M, \sigma_\alpha^2, S, t)$: posterior calibration curve at fixed time knots t_i and global variance σ_α^2.
- $p(g_M | Y, M, S, t)$: another posterior calibration curve (e.g. for radiocarbon). If we can assume that this curve is known *a priori*, then this density will be equal to one.
- $p(\sigma_\alpha^2 | Y, M, S, t)$: posterior density of global variance.
- $p(t | Y, M, S)$: posterior density of calibrated dates arising from Y and M measurements, and stratigraphy S, which also leads to span, first, last or interval estimates.

- $p(S|Y, M)$: posterior probability of the given stratigraphic order S.
- $p(Y)$ and $p(M)$: marginal distributions of Y and M.

3.4.3 Application to Archaeomagnetism

Since the present chapter focuses on archaeomagnetism, I assume that the vector g_M is *a priori* known. Therefore $p(g_M|Y, M, S, t) = 1$ and g_Y will be denoted g for greater simplicity in the rest of the chapter. The measurements Y will represent the hierarchical empirical directions $Y = ((\bar{I}_1, \bar{D}_1), \ldots, (\bar{I}_n, \bar{D}_n))^T$ determined at field level (Section 3.3). Finally, the age measurements M (thermoluminescence, or radiocarbon with dendro-chronological calibration curve $g_M \ldots$), when available, will be used for calculating the prior time t, that is $p(t) = p(t|M, g_M)$. Thus, the Bayesian chronometric Equations 3.6 and 3.7 simplify to

$$p(Y, g, \sigma_\alpha^2, S, t) = p(Y|g, \sigma_\alpha^2, S, t) \times p(g|\sigma_\alpha^2, S, t)$$
$$\times p(\sigma_\alpha^2|S, t) \times p(S|t) \times p(t|M, g_M)$$
$$= p(g|Y, \sigma_\alpha^2, S, t) \times p(\sigma_\alpha^2, t|Y, S) \times p(S|Y) \times p(Y) \quad (3.8)$$

and are illustrated in Figures 3.2 and 3.3.

In what follows, I will develop each of the densities in Equation 3.8 in the case of directional data (based on just inclination and declination), also called spherical data, and illustrate it with the example of the Lübeck site. The univariate (intensity, or inclination alone) and three-dimensional (inclination and declination and intensity) cases are developed in Appendices B and C respectively.

Inclination and Declination Data: the Likelihood (Figure 3.2)

For the spherical case, the chronometric equation takes a particular form because the curve we are searching for draws a path on the surface of the unit sphere, through the inclination and declination data. In practice, direct calculation of the spherical calibration curves seems not to be a feasible proposition, as shown by Jupp and Kent (1987). Hence, taking a slightly different approach, these authors proposed an efficient calculation algorithm based on 'unwrapping' the data $(\bar{I}_i, \bar{D}_i)$ from the surface of the sphere onto the plane (xy), with neither slip nor spin. This unrolling procedure (denoted U for unwrapping) is known to geographers as the zenithal equidistant projection, and in differential geometry as polar normal coordinates or geodesic normal coordinates (Boothby 1975). This leads to the Cartesian coordinates $(x_i, y_i) = (x(t_i), y(t_i)) = U(\bar{I}_i, \bar{D}_i)$, and $(g_{x_i}, g_{y_i}) = (g_x(t_i), g_y(t_i)) = U(g_{I_i}, g_{D_i})$.

In the specific case of directional data with large concentration factors K_{ij} (typically > 50), the error cones $a_{95ij} = t_{\beta F}(r_{ij}) / \sqrt{r_{ij} K_{ij}}$ are conserved with

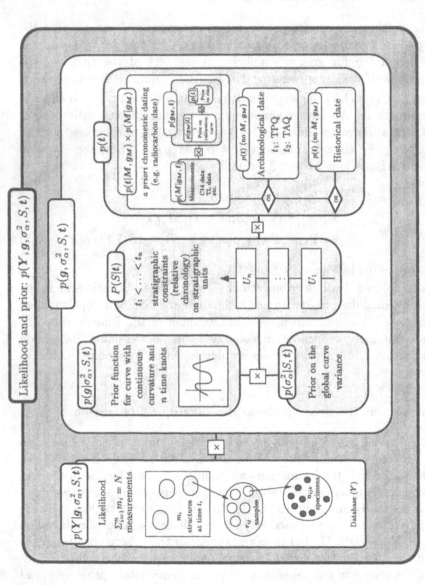

Fig. 3.2. Bayesian chronometric equation applied to archaeomagnetism: joint probability density decomposition into likelihood (measurements) and prior density functions (dating and stratigraphy). × corresponds to multiplication operation. Boxes indicate the hierarchical framework, with posterior densities becoming prior densities for the next level up (see Buck et al. 1996, p. 168).

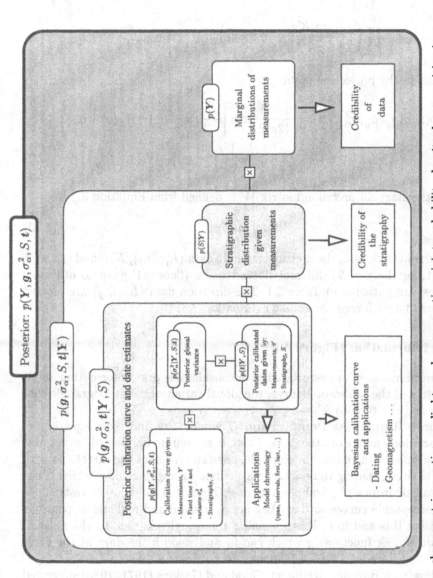

Fig. 3.3. Bayesian chronometric equation applied to archaeomagnetism: joint probability density decomposition into posterior densities (calibration curve, posterior global variance and dates) and marginal densities (stratigraphy and measurements). Boxes indicate the hierarchical framework.

a very high precision during the projection procedure. Then, the likelihood of the directions Y defined on the surface of the sphere

$$p(Y) = \exp\left[\sum_{i=1}^{n} m_i K_{B_i} \left(\cos \bar{I}_i \cos g_{I_i} \cos\left(\bar{D}_i - g_{D_i}\right) + \sin \bar{I}_i \sin g_{I_i}\right)\right]$$
$$\times \prod_{i=1}^{n} \left(\frac{m_i K_{B_i}}{2\pi \sinh(m_i K_{B_i})} \cos \bar{I}_i\right)$$

becomes, in the projection plane (xy),

$$p(x, y | g_x, g_y, \sigma_\alpha^2, S, t) = \frac{\det W}{(2\pi)^n} \exp\left[-\frac{1}{2}\left(x - g_x\right)^T W \left(x - g_x\right)\right]$$
$$\times \exp\left[-\frac{1}{2}\left(y - g_y\right)^T W \left(y - g_y\right)\right] \qquad (3.9)$$

where the diagonal precision matrix W is defined from Equation 3.3 by

$$W_{ii} = m_i K_{B_i}. \qquad (3.10)$$

In the Lübeck case, the archaeomagnetic data $(\bar{I}_{ij}, \bar{D}_{ij})$, K_{ij}, and r_{ij}, with $m_i = 1$, for the $n = 24$ different floors studied (floor '11' gave no utilizable samples) are gathered in Table 3.1. The direction data $(\bar{I}_{ij}, \bar{D}_{ij})$ are plotted in Figure 3.4 with error cones a_{95ij} (Equation 3.5).

Prior Information (Figure 3.2)

The directional data are associated with considerable prior information about the nature of the calibration curve, the global variance, the stratigraphic constrains and the time as follows.

Curve Roughness Penalization. The aim is to find the magnetic secular variation curve from the data. There is an infinite number of curves able to fit or smooth the data. For example, one can represent the directions with pieces of circle, leading to an interpolating but zig-zagging curve which is not very realistic from a physical point of view. Thus, it is necessary to restrict the set of acceptable curves so that they are as natural and realistic as possible. To achieve this and in order to penalize the 'roughness' due to the noise in the data, I seek functions g which can fit and smooth the data at the same time.

Following a Bayesian argument, Good and Gaskins (1971, 1980) suggested a roughness penalization function based on local variation of g. This Bayesian approach (Silverman 1985; Wahba 1990) relies on the idea of defining a prior probability density depending on the curvature of g for a fixed time knot configuration t and for a given smoothing parameter α (Appendix A).

In the spherical case, the prior curve can be defined naturally as

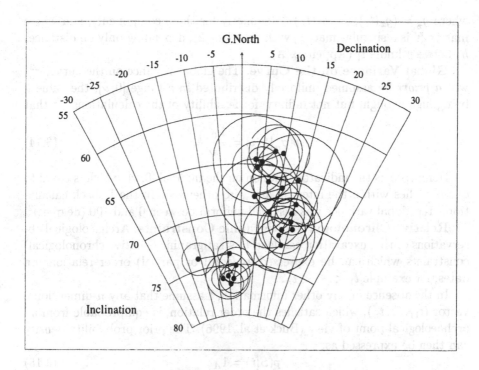

Fig. 3.4. Archaeomagnetic directions of the Lübeck site. Mean directions I and D (black points) are plotted in a stereographic diagram, with error cones at 95% confidence level. Reference directions are connected by straight lines which obey the stratigraphic constraints from the oldest floor (25) to the youngest (1) recovered. G. North indicates the direction of geographic north.

$$p(g|\dots) = C \exp\left(-\frac{\alpha}{2}\int_a^b |\nabla^2 g(t)|^2\, dt\right) \qquad (3.11)$$

which is the spherical analogue to roughness penalization in Euclidean space. The term $\nabla^2 g = (I - gg^T)\,g''$ is the covariant derivative which defines the curvature of g on the surface on the unit sphere (Jupp and Kent 1987). Direct calculation of the spherical roughness penalization is not simple, but using the unrolling procedure U described on page 53 one obtains

$$p(U(g)|\dots) = C \exp\left(\frac{\alpha}{2}\int_a^b g_x''^2 + g_y''^2\, dt\right). \qquad (3.12)$$

If g_x and g_y are natural cubic splines (NCS), then it can be shown that (Green and Silverman 1994, p. 13 and Appendix A)

$$p(U(g)|\dots) = p(g_x, g_y|\sigma_\alpha^2, S, t)$$

$$= \frac{\det^+(\alpha K)}{(2\pi)^{n-2}} \exp\left(-\frac{1}{2}g_x^T(\alpha K)g_x - \frac{1}{2}g_y^T(\alpha K)g_y\right) \qquad (3.13)$$

where $\boldsymbol{g_x} = (g_x(t_1), \ldots, g_x(t_n))^T$ (similarly for the y component), and where matrix $\boldsymbol{K}$ is a singular matrix, with rank $(n-2)$, depending only on distances h_i between knots t_i (Appendix A).

Global Variance on the Curve. The global variance on the curve, σ_α^2, will *a priori* be assumed uniformly distributed in a range $[0, \nu]$, the value ν being high enough, but not infinity, for feasibility of the calculations, so that

$$p(\sigma_\alpha^2|S, t) = \frac{1_{[0,\nu]}}{\nu}. \qquad (3.14)$$

Here, $1_{[0,\nu]}$ is the indicator function on the interval $[0, \nu]$, which is equal to one if σ_α^2 lies within the interval, and zero otherwise. In the Lübeck calculations, the global variance was assumed uniform between 0 and 100 (degrees)^2.

Relative Chronology: Stratigraphic Constraints. Archaeological observations during excavation can lead to stratigraphic (relative chronological) constraints which can be expressed as (total or partial) order relations on dates, for example $t_1 < \cdots < t_n$.

In the absence of any other information, I assume that any n-dimensional vector $(t_1, \ldots, t_n)$, which satisfies the order relation, is equiprobable from an archaeological point of view (Buck et al. 1996). The prior probability density can then be expressed as

$$p(S|t) = 1_A \qquad (3.15)$$

where the set A describes all n-dimensional vectors satisfying the order relation, so that

$$A = \{(t_1, \ldots, t_n) \text{ such that } t_1 < \cdots < t_n\}.$$

In Lübeck, the stratigraphic constraints are obvious (Figure 3.1), the oldest floor being number 25 and the youngest number 1. However, it was not possible to determine whether all the ovens and all their floors had been preserved, how far back in time the sequence of replacements had begun and if there had really been continuous usage. Therefore, no assumption has been made about the rate of 'deposition'. In fact, the span of utilization of each floor will be inferred from the posterior date distributions (see Figures 3.6 and 3.7 and the associated text).

Historical, Archaeological and Chronometric Dating. The problem, then is to date the observations $(\bar{I}_i, \bar{D}_i)$, that is, in the case of archaeomagnetism, to date the remanent magnetization acquired during the last heating of the baked clay.

Some historical documents can provide a very precise date t_h which can be expressed with the Dirac distribution δ, which is equal to one at t_h, and equal to zero otherwise

$$p(t) = \delta_{(t=t_h)}. \qquad (3.16)$$

Most often, however, the prior density, $p(t)$, is defined from a *terminus post quem* (TPQ), the date t_{i1} after which the clay was baked in the case of archaeomagnetism, and a *terminus ante quem* (TAQ), the date t_{i2} before which the clay was baked (Orton 1980, p. 97), provided by archaeology (for example chrono-typology), history or chronometric methods. According to a rather idealized statement, the dating probability is assumed uniform between $t_{i1} = \text{TPQ}$ and $t_{i2} = \text{TAQ}$, and zero outside, if no supplementary information is available (Orton 1980, p. 99) so that

$$p(t) = \frac{1_{[t_{i1},t_{i2}]}}{t_{i2} - t_{i1}}. \tag{3.17}$$

It is possible also to take into account date densities like those provided by chronometric methods. For instance, we can consider normal densities as with TL dates, or more complicated distributions as with calibrated radiocarbon dates. In such situations, for a specific piece of prior information M,

$$p(t|M,g_M) \propto \sqrt{\frac{W_M}{2\pi}} \exp\left[-\frac{1}{2}W_M \left(M - g_M(t)\right)^2\right] \times 1_{(-\infty,0]} \tag{3.18}$$

where

$$W_M = 1/(\sigma_M^2 + \sigma_{g_M}^2).$$

In the case of radiocarbon dating, M is the mean age, σ_M is the error on age (one standard deviation), g_M is the dendrochronological calibration curve (for instance IntCal98) and σ_{g_M} the error on this curve.

For Lübeck, local sources first mention Plot 65 in AD 1301. A transaction describes an existing main building and a baking house, both probably dating from before AD 1284. Extant insurance documents also mention it between 1791 and 1805, when the main building and phase VII oven were constructed, probably around 1800.

Aside from the written sources mentioned above, the styles of bricks used in the construction of the ovens suggest a period of use between the late 13th century, or phase II, and the 18th century, phase VI. However, this dating had a considerable margin of error (50 years or more) and it remained unsatisfactory (Müller 1992), since many of the bricks were re-used.

Thus, it was desirable to confirm the ages using further physical dating methods. ^{14}C-dating and thermoluminescence (TL) dating could in principle be applied, but require suitable material. Unfortunately, only the bottom part of the deposit below floor 25, which consisted of oven clay mixed with rubble of older oven floors, provides organic material (Müller 1992) for ^{14}C-dating. Furthermore clays that were repeatedly heated are not very suited for TL dating and the upper part of the pile may be too young for applying either method.

Prior date intervals are shown in Table 3.1 and Figure 3.5. In Figure 3.5 bold horizontal lines arise from TL dating and each thin horizontal line is deduced from the TPQ of the floor below and the TAQ of the floor above. Prior distributions have been assumed uniform in this study, but in future work I plan to use Equation 3.18 to take account of prior dating distributions provided by C14 and TL.

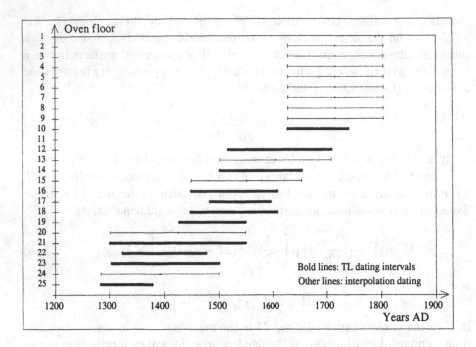

Fig. 3.5. Prior date intervals for the Lübeck site (Germany): bold horizontal lines correspond to TL dating. Each thin horizontal line is a dating interpolation deduced from the TPQ of the floor below and the TAQ of the floor above. Prior distributions are supposed uniform in the intervals.

Posterior Distributions (Figure 3.3)

From the likelihood in Equation 3.9 and priors in Equations 3.13 to 3.18, I make explicit the joint density expression in Equation 3.8 using

$$p(x, y, g_x, g_y, \sigma_\alpha^2, S, t) = \left(\frac{\det W}{(2\pi)^n}\right) \left(\frac{\det^+(aK)}{(2\pi)^{n-2}}\right)$$

$$\times \exp\left[-\frac{1}{2}(x - g_x)^T W (x - g_x) - \frac{1}{2} g_x^T (\alpha K) g_x\right]$$

$$\times \exp\left[-\frac{1}{2}(y - g_y)^T W (y - g_y) - \frac{1}{2} g_y^T (\alpha K) g_y\right]$$

$$\times \frac{1_{[0,\nu]}}{\nu} 1_A \prod_p \delta_{(t=t_h)} \prod_q p(t) \prod_r p(t|M, g_M) \qquad (3.19)$$

where $p + q + r = n$.

A *posteriori*, this joint density becomes the product of three posterior densities and one marginal density, so that

$$p(x, y, g_x, g_y, \sigma_\alpha^2, S, t) = \frac{\det W A^{-1}}{(2\pi)^n}$$

$$\times \exp\left[-\frac{1}{2}(g_x - Ax)^T W A^{-1}(g_x - Ax)\right]$$

$$\times \exp\left[-\frac{1}{2}(g_y - Ay)^T W A^{-1}(g_y - Ay)\right]$$

$$\times \frac{\det^+ W(I - A)}{(2\pi)^{n-2}}$$

$$\times \exp\left[-\frac{1}{2}x^T W(I - A)x - \frac{1}{2}y^T W(I - A)y\right]$$

$$\times p(\sigma_\alpha^2|S, t) \times p(S|t) \times p(t) \qquad (3.20)$$

where I is the unit matrix and where the matrix A (Green and Silverman 1994, pp. 31 and 41) is defined by

$$A = A(\alpha) = (W + \alpha K)^{-1} W. \qquad (3.21)$$

These formulae are a particular case of the general chronometric Equations 3.6 and 3.7, which are adapted to the archaeomagnetic problem (Equation 3.8). Indeed, this conceptualization can be extended to any set of measurements related to time. The functional dependence between the two components (x, y), or (I, D), will be assured through the common smoothing parameter α.

It is easy to consider the geomagnetic intensity, or the geomagnetic inclination alone when one is dealing with sets of bricks and tiles from which declination cannot be inferred. This univariate case is treated in Appendix B. In three-dimensional space, when inclination, declination and intensity are available simultaneously, the general chronometric equation takes a form analogous to Equation 3.20 above (see Appendix C).

In what follows, I discuss how to obtain the posterior calendar dates and the calibration curve on the basis of inclination and declination data.

Posterior Date and Global Variance Distributions. If time knots are randomly distributed according to the prior densities described above, then it is possible to determine the mean time, $E[t]$, and mean global variance, $E[\sigma_\alpha^2]$, the variances, $\mathrm{Var}[t]$ and $\mathrm{Var}[\sigma_\alpha^2]$, and the posterior distributions of the vector t and global variance σ_α^2 themselves

$$p(\sigma_\alpha^2, t|x, y, S) \propto \left(\frac{\det^+ W(I - A)}{(2\pi)^{n-2}}\right) \exp\left[-\frac{1}{2}x^T W(I - A)x\right]$$

$$\times \exp\left[-\frac{1}{2}y^T W(I - A)y\right]$$

$$\times p(\sigma_\alpha^2|S, t) \times p(S|t) \times p(t). \qquad (3.22)$$

This expression provides the way to build a reference chronology combining magnetic measurements, radiocarbon ages (for instance) and stratigraphy.

Random sampling of time and global variance is carried out through a Monte-Carlo method based on the single-component Metropolis–Hastings algorithm, which is itself based on Markov chain theory (so-called MCMC methods, Gilks et al. 1996). This algorithm assures a rapid convergence to the desired precision and allows posterior densities of the dates and the global variance to be estimated simultaneously.

Smoothing Parameter Estimation. The smoothing parameter, α, which represents the 'rate of exchange' between fitting (residual errors) and smoothing (local variations) is estimated by an extension of the generalized cross-validation score $GCV(\alpha)$ (Wahba 1990) to the Bayesian case, so that

$$GCV(\alpha) = \frac{1}{\text{trace}(\boldsymbol{W})} \sum_{i=1}^{n} \left(\int_{t_1,\dots,t_n,\sigma_\alpha^2} W_i \left(\frac{(x_i - (\boldsymbol{Ax})_i)^2 + (y_i - (\boldsymbol{Ay})_i)^2}{(n_F/n)^2} \right)^2 \right.$$

$$\left. \times \, p(\sigma_\alpha^2, t | \boldsymbol{x}, \boldsymbol{y}, S) dt_1 \dots dt_n d\sigma_\alpha^2 \right) \quad (3.23)$$

where n_F is the equivalent degree of freedom for noise (Green and Silverman 1994), defined by

$$n_F = \sum_{i=1}^{n} \left(\int_{t_1,\dots,t_n,\sigma_\alpha^2} (1 - A_{ii}) p(\sigma_\alpha^2, t | \boldsymbol{x}, \boldsymbol{y}, S) dt_1 \dots dt_n d\sigma_\alpha^2 \right). \quad (3.24)$$

The estimate of the smoothing parameter α is the value which minimizes the score $GCV(\alpha)$. Thus, the MCMC calculations have to be performed for a series of values α in order to detect the minimum of GCV.

For Lübeck, calculations were performed by MCMC with $24 \times 20,000$ iterations until precision of the Bayesian time estimates were within two years of one another. After systematic exploration, the best smoothing parameter was found, by generalized cross-validation, to be 5.5×10^{-4}, and this was used to derive the corresponding Bayesian calibration curve (see Figure 3.8 and associated text). The global variance estimate obtained on the spherical curve was $\hat{\sigma}_\alpha^2 = 0.4 \pm 0.3$ (the symbol $\hat{}$ means 'estimate'), which appears close to zero. Posterior date distributions are given in Figure 3.6. From these distributions, it is possible to obtain the highest posterior density interval at 95% for each of the 24 floors in the sequence. As a consequence of the calibration process, we obtain significant improvements of the dating of each of the floors.

Application to Further Chronological Inferences. Following Buck et al. (1991, 1992, 1994, 1996), the posterior densities of dates can also serve to answer different questions concerning beginning or ending or span of a period, or interval between two periods or two floors, order, correlations, and so on. In particular, software developed at Rennes in the specific case of calibration curve estimation is able to produce inferences analagous to those offered by *BCal* (Buck et al. 1999) or *OxCal* (Bronk Ramsey 1995, 1998). The aim in the short term is to make it user-friendly and widely available.

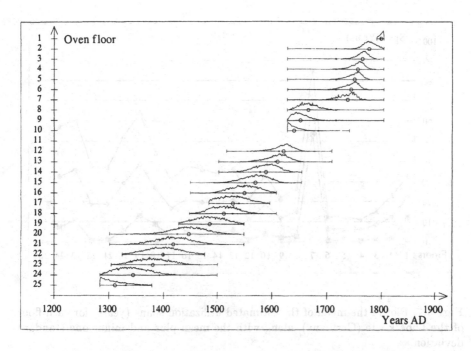

Fig. 3.6. Posterior calibrated dates for each floor of the Lübeck site (Germany): empty circles represent Bayesian mean times. Posterior date distributions have been plotted at yearly intervals, thus the lack of smoothness results from the MCMC sampling. The dating of each floor is significantly improved when compared with the prior.

Thus, in the case of Lübeck, I have been able to establish mean utilization time spans, with their standard deviations, for each floor. These are reported in Figure 3.7. Although the dispersion is large, this clearly shows that the hypothesis of a constant duration of utilization of each floor is not tenable. Thus, durations seem to decrease with time, until the middle of the 17th century. There is a lack of *a priori* well-dated floors for the 18th century, but archaeomagnetic directions for floors 2 to 7 (Figure 3.4) intuitively indicate a trend of events close in time and this is confirmed by Bayesian analysis (Figures 3.6 and 3.8). The larger time interval (about 70 years) between floors 7 and 8 suggests, perhaps, a hiatus or destruction of some floors.

Posterior Curve Distribution. The posterior curve distribution is easily obtained in the spherical case, taking advantage of the linearity of equations. Curve estimation is performed in the plane (xy), leading to a new path $(g_x(t), g_y(t))$ which is rolled from the plane onto the sphere in order to obtain the inclination and declination path $(g_I(t), g_D(t))$. This procedure is iterated until convergence (which is very rapid: about two or three iterations are sufficient). Two different situations can occur which are treated in the next two sections.

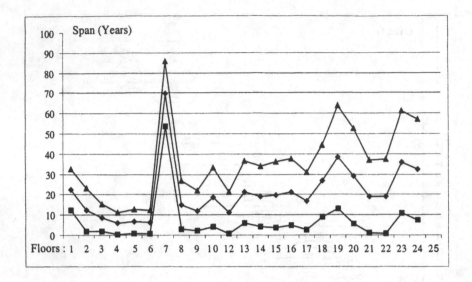

Fig. 3.7. Plot of the mean of the estimated utilization spans (years) for each floor of the Lübeck site (Germany), along with the mean plus and minus one standard deviation.

Fixed Time Knots and Global Variance. For a known time knot configuration t and a known global variance, that is $p(t, \sigma_\alpha^2 | Y, S) = 1$, vectors g_x and g_y will be distributed as

$$p(g_x, g_y | x, y, \sigma_\alpha^2, S, t) = \left(\frac{\det W A^{-1}}{(2\pi)^n} \right)$$

$$\times \exp\left[-\frac{1}{2}(g_x - Ax)^T W A^{-1}(g_x - Ax) \right]$$

$$\times \exp\left[-\frac{1}{2}(g_y - Ay)^T W A^{-1}(g_y - Ay) \right]. \quad (3.25)$$

Vector g_x is multivariate normal with mean Ax and variance–covariance matrix AW^{-1}, and identically for the y component. This density is maximal when

$$g_x = \hat{g}_x = Ax \quad \text{and} \quad g_y = \hat{g}_y = Ay \quad (3.26)$$

where A is the matrix defined in Equation 3.21. The 'influence' matrix A is a convolution matrix which behaves like a variable kernel function whose bandwidth adapts to the density of time knots. Observations close to x_i (and similarly for y_i) contribute most to its estimate, and the speed with which the influence of data points dies away is essentially governed by the local bandwidth, which depends on smoothing parameter α and the local density of points (Silverman 1984). Therefore, this technique is very well adapted to

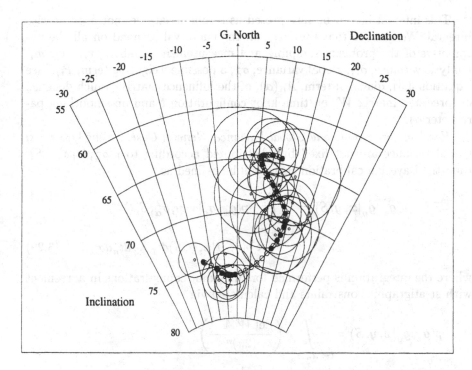

Fig. 3.8. Spherical spline obtained from the Lübeck archaeomagnetic directions. Mean curve is plotted on raw directions of Figure 3.4. Full black points correspond to a Bayesian estimated point, $\hat{g}_{B_i}$, while empty points are spherical interpolation points generated every 10 years by the 'rolling-unrolling' algorithm between sphere surface and plane. G. North indicates the direction of geographic north.

archaeomagnetic data which are often unevenly distributed over time. Both prior and posterior densities are quadratic forms in the function g, and so they correspond to a Gaussian process structure. Since the posterior distribution is also a Gaussian process, the estimate $\hat{g}$ in Equation 3.26, called a 'spline smoother', is the posterior mean as well as being the posterior mode.

One can also determine the highest posterior density interval on the curve, at the 95% level, which will be, because of the framework of the density in Equation 3.25, identical to the confidence interval on the curve obtained from diagonal terms of the influence matrix A (Wahba 1990, pp. 67–71), given by

$$g_x(t_i) = \hat{g}_{x_i} \pm 1.96\sigma_{g_i} \quad \text{and} \quad g_y(t_i) = \hat{g}_{y_i} \pm 1.96\sigma_{g_i} \qquad (3.27)$$

where the error on the curve is

$$\sigma_{g_i} = \sqrt{\left(AW^{-1}\right)_{ii}} = \sqrt{\frac{A_{ii}}{m_i K_{B_i}}}. \qquad (3.28)$$

This interval has to be interpreted as a functional one, not a point-wise interval. We can see that the error on the curve will depend on all the parameters of the problem: sampling and measurement numbers, r_{ij} and m_i, analytic variance, σ_{ij}^2, global variance, σ_α^2, a possible weighting term, P_{ij}, (see Equation 3.3), diagonal term, $A_{ii}(\alpha)$, of the influence matrix (which depends on precision matrix W, on time knot configuration t and on smoothing parameter α).

Random Time-knots and Global Variance: General Case. If time knots and global variance are not fixed but distributed according to $p(\sigma_\alpha^2, t, |x, y, S)$, then the Bayesian calibration curve will be defined as

$$p(g_x, g_y | x, y, S) = \int\limits_{t_1 \ldots t_n, \sigma_\alpha^2} p(g_x, g_y | x, y, \sigma_\alpha^2, S, t)$$
$$\times p(\sigma_\alpha^2, t | x, y, S) dt_1 \ldots dt_n d\sigma_\alpha^2 \qquad (3.29)$$

where the integration is performed for time knot configurations in agreement with stratigraphic constraints and calculated via

$$p(g_x, g_y | x, y, S) = \int\limits_{t_1 \ldots t_n, \sigma_\alpha^2} \left(\frac{\det W A^{-1}}{(2\pi)^n} \right)$$
$$\times \exp \left[-\frac{1}{2}(g_x - Ax)^T W A^{-1}(g_x - Ax) \right]$$
$$\times \exp \left[-\frac{1}{2}(g_y - Ay)^T W A^{-1}(g_y - Ay) \right]$$
$$\times p(\sigma_\alpha^2, t | x, y, S) dt_1 \ldots dt_n d\sigma_\alpha^2. \qquad (3.30)$$

The Bayesian posterior density appears as a continuous 'mixture' of multivariate Gaussian distributions $p(g_x, g_y | x, y, \sigma_\alpha^2, S, t)$ where $p(\sigma_\alpha^2, t | x, y, S)$ is the mixing distribution (Tassi 1992). It is not assured that the posterior density of g remains of Gaussian type, but one can express the Bayesian posterior expectation of g_{x_i} as

$$\hat{g}_{B_{x_i}} = E[g_{x_i}]$$
$$= \int\limits_{g, t, \sigma_\alpha^2} g_{x_i} p(g_x, g_y | x, y, \sigma_\alpha^2, S, t)$$
$$\times p(\sigma_\alpha^2, t | x, y, S) dg_{x_1} \ldots dg_{x_n} dg_{y_1} \ldots dg_{y_n} dt_1 \ldots dt_n d\sigma_\alpha^2$$
$$= \int\limits_{t_1, \ldots, t_n, \sigma_\alpha^2} (Ax)_i \times p(\sigma_\alpha^2, t | x, y, S) dt_1 \ldots dt_n d\sigma_\alpha^2. \qquad (3.31)$$

The estimator $\hat{g}_{B_x}$ is the posterior conditional expectation of the smoother $\hat{g}_x = Ax$ given observations x (see Equation 3.26), y and S, and similarly for the y component. In the same way, one can define a posterior conditional variance on this estimate as

$$\sigma^2_{g_{B_{x_i}}} = \mathrm{E}\left[g^2_{x_i}\right] - \hat{g}^2_{B_{x_i}}$$

$$= \left(\int\limits_{t_1,\ldots,t_n,\sigma^2_\alpha} [(AW^{-1})_{ii} + (Ax)^2_i]\right.$$

$$\left. \times\ p(\sigma^2_\alpha, t | x, y, S) dt_1 \ldots dt_n d\sigma^2_\alpha\right) - \hat{g}^2_{B_{x_i}}. \qquad (3.32)$$

For the y coordinate, we then have

$$\sigma^2_{g_{B_{y_i}}} = \mathrm{E}\left[g^2_{y_i}\right] - \hat{g}^2_{B_{y_i}}.$$

The estimated calibration curve will be plotted using an NCS interpolating function passing through the set of estimated points $\hat{g}_{B_x}(t_i) = \hat{g}_{B_{x_i}}$ and $\hat{g}_{B_y}(t_i) = \hat{g}_{B_{y_i}}$. A 95% highest posterior density interval on the curve (which is no longer a Gaussian distribution), can also be plotted in the same way. Alternatively, one can use a confidence interval defined by the mean error

$$\sigma_{g_{B_i}} = \sqrt{\frac{\sigma^2_{g_{B_{x_i}}} + \sigma^2_{g_{B_{y_i}}}}{2}} = \sqrt{(AW^{-1})_{ii} + \frac{1}{2}\left(\mathrm{Var}\left[(Ax)_i\right] + \mathrm{Var}\left[(Ay)_i\right]\right)}$$

$$(3.33)$$

leading to

$$g_{B_x}(t_i) = \hat{g}_{B_{x_i}} \pm 1.96\sigma_{g_{B_i}} \text{ and } g_{B_y}(t_i) = \hat{g}_{B_{y_i}} \pm 1.96\sigma_{g_{B_i}}. \qquad (3.34)$$

The calibration curve on inclination and declination is obtained by rolling the parametric curve from the plane (xy) onto the sphere according to the inverse projection $(\hat{g}_{B_I}(t), \hat{g}_{B_D}(t)) = U^{-1}(\hat{g}_{B_x}(t), \hat{g}_{B_y}(t))$. The confidence intervals on the two Bayesian marginal curves on inclination and declination are then

$$g_{B_I}(t_i) = \hat{g}_{B_{I_i}} \pm 1.96\sigma_{g_{B_i}} \text{ and } g_{B_D}(t_i) = \hat{g}_{B_{D_i}} \pm 1.96\sigma_{g_{B_i}} / \cos\hat{g}_{B_{I_i}}. \qquad (3.35)$$

For Lübeck, the spherical spline obtained on the sphere is plotted in Figure 3.8 (mean curve on raw data) and in Figure 3.9 (mean curve and associated error circles $a_{95i} = 2.45\sigma_{g_{B_i}}$ at 95% level). Marginal curves on inclination and declination are plotted in Figures 3.10 and 3.11, with error envelope at 95%. Black circles along the curves correspond to Bayesian time estimations while empty circles, displayed for information, are spherical interpolating points generated by the 'rolling–unrolling' algorithm between sphere surface and projection plane. The inclination shows a minimum in the 14th century and increases then by more than 10°. Declination shows a local minimum around AD 1400 followed by a maximum in the 17th century, after that declination moves about 30° to the west. Correlation with the historical magnetic record of direct directional readings (Schnepp et al. 2003) suggests that the age estimate for the upper ten floors was too young and must date from the end of the 16th to the mid-18th century. For the lowermost 14 floors dating is reliable and provides a secular variation curve for Germany.

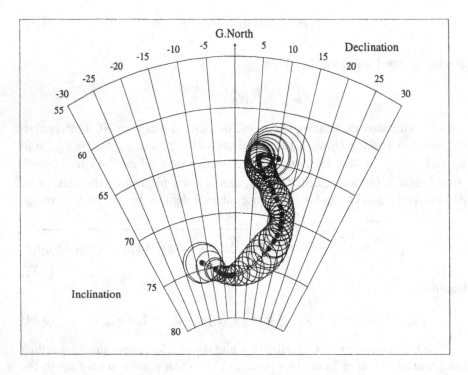

Fig. 3.9. Spherical spline obtained from the Lübeck archaeomagnetic directions. Mean curve (black points) and HPD circles at 95% level. Circle sizes are clearly related to density of directions on the sphere as well as distances between Bayesian estimated points $\hat{g}_{B_i}$. G. North indicates the direction of geographic north.

Credibility of Stratigraphy and Data. The posterior distribution on S and the marginal distribution on x and y can provide some information about the credibility of the posterior inferences. A high posterior probability $p(S|x, y)$ means that S agrees well with the data and the other prior information of the model. On the other hand, one can calculate the conditional marginal density $p(x_i, y_i | x_1, y_1, \ldots, x_{i-1}, y_{i-1}, x_{i+1}, y_{i+1}, \ldots, x_n, y_n)$ and define an HPD area, or credibility area (Ulrych et al. 1992), for the variables (x_i, y_i). This is a way to detect some outlying data which can be aberrant (due to systematic error) or not correctly dated.

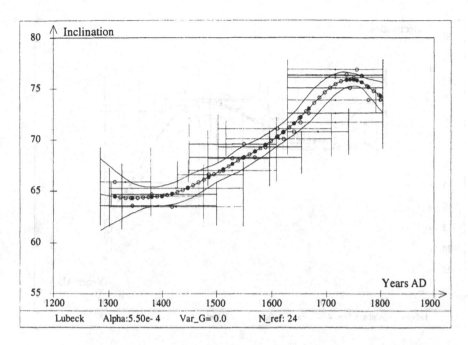

Fig. 3.10. Marginal mean curve on inclination with functional HPD envelope at 95% level. Black circles along with the curve correspond to Bayesian time estimations while empty circles are spherical interpolating points generated by the 'rolling–unrolling' algorithm. Prior (uniform) distributions for the dates of the 24 oven floors are plotted as horizontal bars, and errors on the inclination data are represented by vertical bars at each end point of the horizontal bars.

3.5 Archaeomagnetic Dating of a New Sample

Knowing the calibration curve for a given area, it is possible to date a new structure of unknown date in the same area. Archaeomagnetic date calculation is implemented in a Bayesian context, in a similar way to radiocarbon calibration (Pazdur and Michczynska 1980; Stuiver and Reimer 1989; van der Plicht and Mook 1989; Michczynska et al. 1990; Dehling and van der Plicht 1993; van der Plicht 1993). Here, age is replaced by inclination and declination, and the calibration function (for instance IntCal98) is replaced by geomagnetic secular variation curves deduced from previous calculations or from archaeomagnetic literature. In the spherical case, the direction $Z = (I, D)$ one wants to date is projected onto the plane (xy) according to the unrolling procedure described on page 53. There, the projected direction (x_Z, y_Z) can be dated with the Bayesian calibration curves $\hat{g}_{B_x}(t)$ and $\hat{g}_{B_y}(t)$.

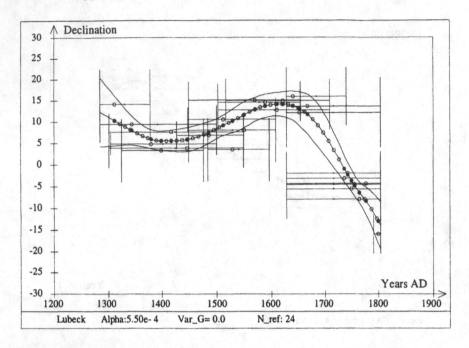

Fig. 3.11. Marginal mean curve on declination with functional HPD envelope at 95% level. See caption to Figure 3.10.

3.5.1 Likelihood

The likelihood is presented for the component x, the reasoning being the same for the y component. The observation x_Z, which is a statistical estimation of a true (but unknown) value $g_{B_x}(t)$, is supposed normally distributed at a fixed time t, with true mean $g_{B_x}(t)$, and empirical error $\sigma_Z^2 = a_{95Z}/2.45$, that is $x_Z \sim N(g_{B_x}(t), \sigma_Z^2)$. The unknown geomagnetic secular variation curve $g_{B_x}(t)$ will be supposed normally distributed with mean $\hat{g}_{B_x}(t) = \mathrm{E}[g_{B_x}(t)|t]$ and variance $\sigma_{g_B}^2$ given by Equation 3.33 (Equation 3.28 being a particular case of Equation 3.33, when time knots are fixed), that is $g_{B_x}(t) \sim N(\hat{g}_{B_x}(t), \sigma_{g_B}^2)$, whose density is denoted $p(g_{B_x}|\hat{g}_{B_x}, t)$.

The likelihood, or conditional probability density of the observation x_Z, denoted $p(x_Z|\hat{g}_{B_x}, t)$, is linked to time t via the formula

$$p(x_Z|\hat{g}_{B_x}, t) = \int_{-\infty}^{+\infty} p(x_Z|g_{B_x}, \hat{g}_{B_x}, t)p(g_{B_x}|\hat{g}_{B_x}, t)dg_{B_x}$$

where the unknown variable g_{B_x} is eliminated after an integration process. We deduce that x_Z will be normally distributed with mean $\hat{g}_{B_x}$ and variance $(\sigma_Z^2 + \sigma_{g_B}^2)$. Dependence of x_Z on t occurs through the estimated function $\hat{g}_{B_x}$, given by Equation 3.31, and through the new variance which takes into account errors in both the curve and laboratory measurements.

3.5.2 Prior Probability Density on a Date

All calendar past dates are *a priori* equiprobable for the artefact to be dated. Consequently

$$p(t|\hat{g}_{B_x}) = 1_{(-\infty,0]}.$$

3.5.3 Posterior Probability Density: 'Calibrated' Date

The posterior probability density will be calculated using the projected direction (x_Z, y_Z) and parametric curves $\hat{g}_{B_x}$ and $\hat{g}_{B_y}$ in the zenithal projection plane as

$$p(t|x_Z, y_Z, \hat{g}_{B_x}, \hat{g}_{B_y}) \propto \frac{W_{Zg}}{2\pi} \exp\left[-\frac{1}{2}W_{Zg}(x_Z - \hat{g}_{B_x}(t))^2\right]$$

$$\times \exp\left[-\frac{1}{2}W_{Zg}(y_Z - \hat{g}_{B_y}(t))^2\right] \times 1_{(-\infty,0]} \quad (3.36)$$

where

$$W_{Zg} = 1/(\sigma_Z^2 + \sigma_{g_B}^2).$$

The variance on the projected calibration curve $\sigma_{g_B}^2$ is determined at time t thanks to a natural cubic spline interpolation between values given by Equation 3.33 at each time knot t_i.

The framework for archaeomagnetic dating summarized by Equation 3.36 does not mean that the two variables x_Z and y_Z are independent. In fact, they are linked together through the smoothing parameter α estimated on the sphere. This general dating technique applied to scalar or vector data can be compared with an alternative technique recently proposed by Le Goff et al. (2002) and based on the use of the test suggested by McFadden and McElhinny (1990).

As a formal example, in Figure 3.12 I show one such dating process for a site with direction $I = 70°$ and $D = 14°$ East, angular error $a_{95Z} = 1°$, and Bayesian marginal curves on inclination $\hat{g}_{B_I}$ and declination $\hat{g}_{B_D}$. The posterior 95% HPD interval obtained using just the inclination is AD [1563, 1635] (Figure 3.12a). Using just the declination, I obtain two intervals AD [1284, 1318] and [1537, 1682] (Figure 3.12b). The combination of the two parameters allows me to select one of the two solutions given by the declination, while slightly reducing the final dating interval to AD [1567, 1633] (Figure 3.12c). Thus, the error range at 95% is about 70 years for this part of the calibration curve. The same result is obtained if the calculation is directly performed in the plane (xy) with curves $\hat{g}_{B_x}$ and $\hat{g}_{B_y}$.

3.6 Conclusion

I have proposed a general coherent solution to the calibration curve building problem, illustrated in this chapter in the archaeomagnetic dating case. In

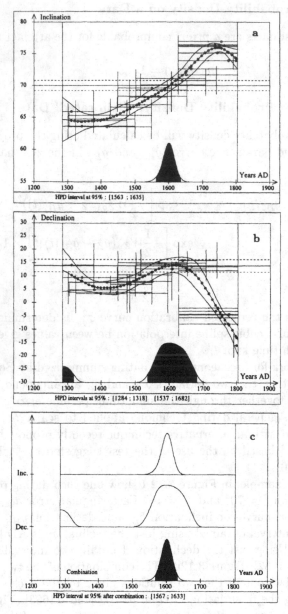

Fig. 3.12. Posterior dates for an archaeological site whose archaeomagnetic direction is characterized by: $I = 70°$ and $D = 14°$ East with angular error $a_{95} = 1°$ estimated using the Lübeck calibration data. a) calibrated date interval obtained with inclination curve: 95% HPD AD [1563, 1635], b) calibrated date intervals obtained with declination: 95% HPD AD [1284, 1318] and [1537, 1682], c) calibrated date interval obtained by combining inclination and declination: 95% HPD AD [1567, 1633].

fact, this modelling can be extended to any other types of data (scalar or vector): tree-ring widths in dendrochronology, depth in a sediment deposit (such as in a peat bog) or proportion of a specific feature. All hypotheses are explicitly formulated with the help of a probabilistic approach based on Bayesian concepts. In particular, emphasis is put on relationships between prior dating and functional parameters on the one hand, and observational data on the other. One of the most important contributions of this modelling is to allow definition of strict rules for the use of prior and posterior dating. In particular, dating of each *a priori* dated reference structure used for calibration curve building is *a posteriori* improved with the Bayesian calibration curve calculation. Reference chronology and calibration curves are products of the same global statistical approach. Once established, this reference curve can also serve to date other independent structures which are not included in the curve building, using the dating process described in Section 3.5.

Acknowledgements

I am very indebted to E. Schnepp for having put Lübeck archaeomagnetic data at my disposal. Dr. C. Buck and Prof. A. O'Hagan (Department of Probability and Statistics, University of Sheffield, UK) and the two reviewers are highly thanked for their critical and constructive comments. I also thank Ph. Dufresne (Rennes Laboratory) for his help with figure drawing.

Appendix A: Prior Density on the Curve

A way to measure how 'rough' or 'wiggly' the curve g is, is to calculate the integral of the second derivative squared, taken on an interval $[a, b]$ such that $a < t_1 < \cdots < t_n < b$

$$\int_a^b g''(t)^2 dt.$$

This implies that g is twice differentiable on this interval. Thus the prior density can be defined as:

$$p(g) = C \exp \left[-\frac{\alpha}{2} \int_a^b g''(t)^2 dt \right] \qquad (3.37)$$

where C is a normalization factor that will be detailed in Equation 3.39, and where α is a smoothing parameter. The greater the smoothing parameter α, the greater the weight that is placed on weak roughness functions, and thus the greater the smoothing effect. If g is a natural cubic spline (NCS), then it can be shown that (Green and Silverman 1994, p. 13)

$$\int_a^b g''(t)^2 dt = g^T K g \qquad (3.38)$$

where matrix K is a singular matrix, with rank $(n-2)$, depending only on distances h_i between knots t_i, that is $h_i = t_{i+1} - t_i$. It has two null eigenvalues which correspond to constant and linear functions g. Thus, the prior density is 'partially improper' in that it is invariant under the addition of a constant or linear function to g, and is multivariate normal with variance matrix $(\alpha K)^{-1}$. This density has infinite variance for two eigenvalues, but it can be normalized provided that the determinant of K is defined on non-zero eigenvalues, hence the notation $\det^+$. The prior density can then be written

$$p(g|\sigma_\alpha^2, S, t) = C \exp\left[-\frac{\alpha}{2}\int_a^b g''(t)^2 dt\right]$$

$$= \sqrt{\frac{\det^+(\alpha K)}{(2\pi)^{n-2}}} \exp\left[-\frac{1}{2}g^T(\alpha K)g\right]. \tag{3.39}$$

Although it may appear so, NCS functions are not chosen *a priori*, but these functions are the unique minimizers of the roughness measure over the class of all continuous and twice differentiable functions on $[a, b]$, which interpolate $g(t_i) = g_i$ (Green and Silverman 1994, pp. 15 and 16). Thus, in the roughness penalization approach, NCS arises as a mathematical consequence of the choice of the roughness penalty. This choice means that the physical phenomena under study vary in a smooth way with continuous first and second derivatives. Magnetic data continuously recorded in observatories (e.g. Alexandrescu et al. 1996) can, at a first approximation, justify such a choice.

Appendix B: Modelling Field Intensity (Univariate Case)

B.1 Hierarchical Modelling

Using the Normal statistics at each level according to a hierarchical Bayesian approach, the relationship between any single magnetic property such as the measured intensities, F_{ijk}, at sample level and the hierarchical empirical intensity, $\bar{F}_i$, at field level is established through an integration with respect to the unknown variable F_{ij} (Lanos et al. submitted). This leads to the Bayesian hierarchical (empirical, or observed) intensity

$$\bar{F}_i = \sum_{j=1}^{m_i} P_{ij}\bar{F}_{ij} \bigg/ \sum_j P_{ij} \tag{3.40}$$

where

$$\bar{F}_{ij} = \frac{1}{r_{ij}}\sum_{k=1}^{r_{ij}} F_{ijk}$$

and where the number of samples is r_{ij} and the number of structures is m_i.

The distribution of the empirical mean intensity, $\bar{F}_i$, at field level is a hierarchical Gaussian statistic around the unknown intensity $g_i = g_{F_i}$ of the magnetic field

$$p\left(\bar{F}_i\right) = \frac{\sqrt{m_i}}{\sigma_{B_i}\sqrt{2\pi}} \exp\left[-\frac{1}{2}m_i\left(\frac{\bar{F}_i - g_{F_i}}{\sigma_{B_i}}\right)^2\right] \qquad (3.41)$$

where $\sigma_{B_i}^2$ is the variance which characterizes the dispersion of the empirical mean intensity around the true (unknown) mean intensity. This variance appears as the inverse of a harmonic mean of variances σ_i^2 and σ_{ij}^2 which are defined at field and structure levels respectively (Lanos et al. submitted) so that

$$\frac{1}{\sigma_{B_i}^2} = \frac{1}{m_i}\sum_{j=1}^{m_i} P_{ij}\left(\sigma_i^2 + \frac{P_{ij}\sigma_{ij}^2}{r_{ij}}\right)^{-1}. \qquad (3.42)$$

The variance σ_{ij}^2, is estimated from the unbiased empirical variance

$$\hat{\sigma}_{ij}^2 = \frac{1}{(r_{ij}-1)}\sum_{k=1}^{r_{ij}}\left(F_{ijk} - \bar{F}_{ij}\right)^2. \qquad (3.43)$$

The unknown global variance, σ_i^2, will be supposed constant over the entire length of the calibration curve and thus equal to a global variance σ_α^2 which will be estimated during curve estimation (this simplification is unnecessary if there are enough data for inferring σ_i^2 at each individual time knot t_i). The global variance corresponds to the residual variance defined by Green and Silverman (1994, p. 38) and will be estimated, following the Bayesian approach, by assuming some vague prior knowledge about it (see Equation 3.14 and associated text).

An error interval at the 95% level around the empirical mean is determined by

$$E_{95i} = 1.96/\sqrt{m_i/\sigma_{B_i}^2}. \qquad (3.44)$$

B.2 Likelihood

In the case of n independent normal univariate observations $\bar{F}_i$, represented by vector $\boldsymbol{F} = (\bar{F}_1, \ldots, \bar{F}_n)^T$, and defined at n time knots t_i and following the Bayesian hierarchical model, the joint probability density function, or likelihood, will be

$$p(\boldsymbol{F}|\boldsymbol{g}, \sigma_\alpha^2, S, t) = \sqrt{\frac{\det \boldsymbol{W}}{(2\pi)^n}} \exp\left[-\frac{1}{2}(\boldsymbol{F} - \boldsymbol{g})^T \boldsymbol{W}(\boldsymbol{F} - \boldsymbol{g})\right] \qquad (3.45)$$

where terms of the diagonal 'precision' matrix $\boldsymbol{W}$ (inverse squared errors) are given by $W_i = m_i/\sigma_{B_i}^2$.

B.3 Prior on the Curve

From Equation 3.13, I deduce the form of the prior density on the curve in the univariate case

$$p(g|\sigma_\alpha^2, S, t) = \sqrt{\frac{\det^+(\alpha K)}{(2\pi)^{n-2}}} \exp\left[-\frac{1}{2} g^T (\alpha K) g\right]. \qquad (3.46)$$

B.4 The Posterior

Posterior distributions are immediately deduced from Equations 3.19 and 3.20, considering only one variable. In the case of fixed time knots, the error on the curve will be defined by

$$\sigma_{g_i} = \sqrt{(AW^{-1})_{ii}} = \sqrt{\frac{A_{ii}\sigma_{B_i}^2}{m_i}} \qquad (3.47)$$

where A is the influence matrix defined in Equation 3.21. In the case of random time knots, the errors on the calibration curve are given in Equations 3.29 to 3.35.

B.5 The Calibrated Date

The calibrated date will be obtained from the observation intensity F_Z with empirical error σ_Z (see Section 3.5)

$$p(t|F_Z, \hat{g}_{B_F}) \propto \frac{W_{F_{Z_g}}}{2\pi} \exp\left[-\frac{1}{2} W_{F_{Z_g}} (F_Z - \hat{g}_{B_F}(t))^2\right] \qquad (3.48)$$

where

$$W_{F_{Z_g}} = 1/\left(\sigma_Z^2 + \sigma_{g_{B_F}}^2\right).$$

The calibration curve, $\hat{g}_{B_F}$, is given by Equation 3.31 and the curve variance, $\sigma_{g_{B_F}}^2$, by Equation 3.32.

Appendix C: Modelling Intensity, Declination and Inclination (Three-Dimensional Case)

C.1 Hierarchical Modelling

The hierarchical modelling can be extended to the vector case in three-dimensional space. Thus, the magnetic field (inclination, declination and intensity) at time t_i can be decomposed as independent normal components g_x, g_y, and g_z in the Cartesian coordinate system (xyz), according to an orthogonal projection on axes x, y and z. For instance, at field level,

$$x_i = \bar{F}_i \cos \bar{I}_i \cos \bar{D}_i \qquad y_i = \bar{F}_i \cos \bar{I}_i \sin \bar{D}_i \qquad z_i = \bar{F}_i \sin \bar{I}_i$$

$$g_{x_i} = g_{F_i} \cos g_{I_i} \cos g_{D_i} \qquad g_{y_i} = g_{F_i} \cos g_{I_i} \sin g_{D_i} \qquad g_{z_i} = g_{F_i} \sin g_{I_i}.$$

The precision diagonal matrices $\boldsymbol{W_x}$, $\boldsymbol{W_y}$ and $\boldsymbol{W_z}$ are determined using the 'delta' approximation lemma (Mardia 1972, p. 111) and knowing that the intensity, $\bar{F}_i$, is normally distributed and that the direction $(\bar{I}_i, \bar{D}_i)$ is Fisher distributed,

$$W_{x_i} = \frac{m_i}{V_{x_i}} \qquad W_{y_i} = \frac{m_i}{V_{y_i}} \qquad W_{z_i} = \frac{m_i}{V_{z_i}} \qquad (3.49)$$

with $\quad V_{x_i} \approx \sigma_{B_i}^2 \left(\cos^2 \bar{I}_i \cos^2 \bar{D}_i \right) + \dfrac{\bar{F}_i^2}{K_{B_i}} \left(\sin^2 \bar{I}_i \cos^2 \bar{D}_i + \sin^2 \bar{D}_i \right)$

and $\quad V_{y_i} \approx \sigma_{B_i}^2 \left(\cos^2 \bar{I}_i \sin^2 \bar{D}_i \right) + \dfrac{\bar{F}_i^2}{K_{B_i}} \left(\sin^2 \bar{I}_i \sin^2 \bar{D}_i + \cos^2 \bar{D}_i \right)$

and $\quad V_{z_i} \approx \sigma_{B_i}^2 \left(\sin^2 \bar{I}_i \right) + \dfrac{\bar{F}_i^2}{K_{B_i}} \left(\cos^2 \bar{I}_i \right)$

where $1/K_{B_i}$ and $\sigma_{B_i}^2$ are given by Equations 3.3 and 3.42 respectively.

If $\bar{F}_i^2 / K_{B_i} = \sigma_{B_i}^2 = \sigma^2$, then $V_{x_i} = V_{y_i} = V_{z_i} = \sigma^2$. This particular result corresponds to a Gaussian distribution where the magnetic field is characterized by a mean vector, represented by g_{x_i}, g_{y_i} and g_{z_i}, together with a dispersion vector, with isotropic variance σ^2 about the end point of the vector (as considered by Love and Constable 2003). In this case, the variance σ^2 can be estimated by

$$\hat{\sigma}^2 = \frac{1}{3} \left(\sigma_{B_i}^2 + 2 \frac{\bar{F}_i^2}{K_{B_i}} \right). \qquad (3.50)$$

C.2 Likelihood

The likelihood for component x becomes

$$p(x|g_x, \sigma_\alpha^2, S, t) = \sqrt{\frac{\det \boldsymbol{W_x}}{(2\pi)^n}} \exp \left[-\frac{1}{2}(x - g_x)^T \boldsymbol{W_x}(x - g_x) \right] \qquad (3.51)$$

where terms of the diagonal 'precision' matrix $\boldsymbol{W_x}$ (inverse squared errors) are given in Section C.1. The same formulation is used for components y and z.

C.3 Prior on the Curve

The prior density on the curve will be defined, for component x thus

$$p(g_x|\sigma_\alpha^2, S, t) = \sqrt{\frac{\det^+(\alpha K)}{(2\pi)^{n-2}}} \exp\left[-\frac{1}{2}g_x^T(\alpha K)g_x\right]. \qquad (3.52)$$

However surprising it may be, this three-dimensional case is simpler to solve than the spherical case with inclination and declination, and relies directly on using the univariate penalized spline algorithm. We notice that the dependence between the three geomagnetic variables will be assured via the common smoothing parameter α, errors on time (if random time knots) and global variance σ_α^2.

C.4 The Posterior

Posterior distributions are deduced from Equations 3.19 and 3.20. In the three-dimensional case, the errors on the three calibration curves, for fixed time knots, are defined by

$$\sigma_{g_{x_i}} = \sqrt{A_{x_{ii}}(\alpha)/W_{x_i}} \quad \sigma_{g_{y_i}} = \sqrt{A_{y_{ii}}(\alpha)/W_{y_i}} \quad \sigma_{g_{z_i}} = \sqrt{A_{z_{ii}}(\alpha)/W_{z_i}}$$
$$(3.53)$$

where precision matrices W_x, W_y and W_z are given in Equation 3.49 and influence matrices A_x, A_y and A_z are defined by Equation 3.21, W being replaced by W_x, W_y and W_z respectively. In the case of random time knots, the errors on the three calibration curves are given by Equations 3.29 to 3.35.

C.5 The Calibrated Date

The use of two or three geomagnetic parameters allows the dating precision to be increased and often reduces the number of posterior chronological intervals obtained. The geomagnetic measurement vector of unknown date (inclination I_Z, declination D_Z and intensity F_Z) is expressed in geocentric Cartesian coordinates as x_Z, y_Z and z_Z where

$$x_Z = F_Z \cos I_Z \cos D_Z \qquad y_Z = F_Z \cos I_Z \sin D_Z \qquad z_Z = F_Z \sin I_Z$$

with the associated errors σ_{x_Z}, σ_{y_Z} and σ_{z_Z}. The estimated reference curve characteristics are $\hat{g}_{B_x}$, $\hat{g}_{B_y}$ and $\hat{g}_{B_z}$. Then, the posterior probability density of the date will readily be obtained from

$$p(t|x_Z, y_Z, z_Z, \hat{g}_{B_x}, \hat{g}_{B_y}, \hat{g}_{B_z}) \propto \sqrt{\frac{W_{xz_g}}{2\pi}} \exp\left[-\frac{1}{2}W_{xz_g}\left(x_Z - \hat{g}_{B_x}(t)\right)^2\right]$$

$$\times \sqrt{\frac{W_{yz_g}}{2\pi}} \exp\left[-\frac{1}{2}W_{yz_g}\left(y_Z - \hat{g}_{B_y}(t)\right)^2\right]$$

$$\times \sqrt{\frac{W_{zz_g}}{2\pi}} \exp\left[-\frac{1}{2}W_{zz_g}\left(z_Z - \hat{g}_{B_z}(t)\right)^2\right]$$

$$\times 1_{(-\infty, 0]} \tag{3.54}$$

where

$$W_{xz_g} = 1/\left(\sigma_{xz}^2 + \sigma_{g_{B_x}}^2\right)$$

and similarly for W_{yz_g} and W_{zz_g}.

References

Alexandrescu, M., Courtillot, V. and Mouël, J.-L. L. (1996). Geomagnetic field direction in Paris since the mid-sixteenth century. *Physics of the Earth and Planetary Interiors*, **98**, 321–360.

Batt, C. M. (1997). The British archaeomagnetic calibration curve: an objective treatment. *Archaeometry*, **39**, 163–168.

Boothby, W. M. (1975). *An introduction to differentiable manifolds and Riemannian geometry*. Academic Press, New York.

Bronk Ramsey, C. (1995). Radiocarbon calibration and analysis of stratigraphy: the OxCal program. *Radiocarbon*, **37**, 425–430.

Bronk Ramsey, C. (1998). The role of statistical methods in the interpretation of radiocarbon dates. In J. Evin, C. Oberlin, J. P. Daugas and J. F. Salles (eds.), *¹⁴C et archéologie: 3ème congrès international, Lyon, 6–10 avril 1998*. Mémoires de la Société Préhistorique Française Tome XXVI et Supplément 1999 de la Revue d'Archéometrie, 83–86.

Buck, C. E., Cavanagh, W. G. and Litton, C. D. (1996). *Bayesian approach to interpreting archaeological data*. John Wiley, Chichester.

Buck, C. E., Christen, J. A. and James, G. N. (1999). BCal: an on-line Bayesian radiocarbon calibration tool. *Internet Archaeology*, **7**.
URL http://intarch.ac.uk/journal/issue7/buck/

Buck, C. E., Kenworthy, J. B., Litton, C. D. and Smith, A. F. M. (1991). Combining archaeological and radiocarbon information: a Bayesian approach to calibration. *Antiquity*, **05**, 808–821.

Buck, C. E., Litton, C. D. and Shennan, S. J. (1994). A case study in combining radiocarbon and archaeological information: the early Bronze Age settlement of St. Veit-Klinglberg, Land Salzburg, Austria. *Germania*, **72**, 427–447.

Buck, C. E., Litton, C. D. and Smith, A. F. M. (1992). Calibration of radiocarbon results pertaining to related archaeological events. *Journal of Archaeological Science*, **19**, 497–512.

Bucur, I. (1994). The direction of the terrestrial magnetic field in France during the last 21 centuries. Recent progress. *Physics of the Earth and Planetary Interiors*, **87**, 95–109.

Chauvin, A., Garcia, Y., Lanos, P. and Laubenheimer, F. (2000). Palaeointensity of the geomagnetic field recovered on archaeomagnetic sites from France. *Physics of the Earth and Planetary Interiors*, **120**, 111–136.

Dehling, H. and van der Plicht, J. (1993). Statistical problems in calibrating radiocarbon dates. *Radiocarbon*, **35**, 239–244.

Fisher, R. A. (1953). Dispersion on a sphere. *Proceedings of Royal Society Series A*, **217**, 295.

Gallet, Y., Genevey, A. and Le Goff, M. (2002). Three millennia of directional variation of the Earth's magnetic field in western Europe as revealed by archaeological artefacts. *Physics of the Earth and Planetary Interiors*, **131**, 81–89.

Gilks, W., Richardson, S. and Spiegelhalter, D. (eds.) (1996). *Markov chain Monte Carlo in practice*. Chapman and Hall, London.

Good, I. J. and Gaskins, R. A. (1971). Nonparametric roughness penalties for probability densities. *Biometrika*, **58**, 255–277.

Good, I. J. and Gaskins, R. A. (1980). Density estimation and bump-hunting by the penalized maximum likelihood method exemplified by scattering and meteorite data. *Journal of the American Statistical Association*, **75**, 42–73.

Green, P. J. and Silverman, B. W. (1994). *Nonparametric regression and generalized linear models, a roughness penalty approach*. Chapman and Hall, London.

Jupp, P. E. and Kent, J. T. (1987). Fitting smooth paths to spherical data. *Applied Statistics*, **36**, 34–46.

Kovacheva, M. (1997). Archaeomagnetic database from Bulgaria: the last 8000 years. *Physics of the Earth and Planetary Interiors*, **102**, 145–151.

Kovacheva, M., Jordanova, N. and Karloukovski, V. (1998). Geomagnetic field variations as determined from Bulgarian archaeomagnetic data. Part II: The last 8000 years. *Surveys in Geophysics*, **19**, 431–460.

Kovacheva, M. and Toshkov, A. (1994). Geomagnetic field variations as determined from Bulgarian archaeomagnetic data. Part I: The last 2000 years AD. *Surveys in Geophysics*, **15**, 673–701.

Lanos, P. (2001). L'approche bayésienne en chronométrie: application à l'archéomagnétisme. In J. N. Barrandon, P. Guibert and V. Michel (eds.), *Datation, XXIe rencontres internationales d'archéologie et d'histoire d'Antibes*. APDCA, Antibes, 113–139.

Lanos, P., Kovacheva, M. and Chauvin, A. (1999). Archaeomagnetism, methodology and applications: implementation and practice of the archaeomagnetic method in France and Bulgaria. *European Journal of Archaeology*, **2**, 365–392.

Lanos, P., Le Goff, M., Kovacheva, M. and Schnepp, E. (submitted). Archaeomagnetic reference curves, part I: errors from sampling to databases and curve estimation by moving average technique. *Geophysical Journal International*.

Le Goff, M., Gallet, Y., Genevey, A. and Warmé, N. (2002). On archaeomagnetic secular variation curves and archaeomagnetic dating. *Physics of the Earth and Planetary Interiors*, **134**, 203–211.

Love, J. J. and Constable, C. G. (2003). Gaussian statistics for paleomagnetic vectors. *Geophysical Journal International*, **152**, 515–565.

Mardia, K. V. (1972). *Statistics of directional data*. Probability and Mathematical Statistics Series. Academic Press, London.

Marton, P. (1996). Archaeomagnetic directions: the Hungarian calibration curve. In A. Morris and D. Tarling (eds.), *Palaeomagnetism and tectonics of the Mediterranean region*, Geological Society Special Publication, vol. 105, 385–399.

McFadden, P. L. and McElhinny, M. (1990). Classification of the reversal test in palaeomagnetism. *Geophysical Journal International*, **103**, 725–729.

Michczynska, D. J., Pazdur, M. F. and Walanus, A. (1990). Bayesian approach to probabilistic calibration of radiocarbon ages. In W. G. Mook and H. T. Waterbolk (eds.), ^{14}C *and archaeology: proceedings of the second international symposium, Groningen 1987*. PACT, Strasbourg, Journal of the European Study Group on Physical, Chemical and Mathematical Techniques Applied to Archaeology, Council of Europe, **29**, 69–79.

Müller, U. (1992). Eine gewerbliche Bäckerei in Lübeck vom 13. bis zum 20. Jahrhundert. Ergebnisse der Grabung Mühlenstraße. *Lübecker Schriften zur Archäologie und Kulturgeschichte*, **22**, 123–143.

Orton, C. (1980). *Mathematics in archaeology*. Collins, London.

Pazdur, M. F. and Michczynska, D. J. (1980). Improvement of the procedure for probabilistic calibration of radiocarbon dates. *Radiocarbon*, **31**, 824–832.

Schnepp, E. and Pucher, R. (1998). Preliminary archaeomagnetic results from a floor sequence of a bread kiln in Lübeck (Germany). *Studia geophysica et geodaetica*, **42**, 1–11.

Schnepp, E., Pucher, R., Goedicke, C., Manzano, A., Müller, U. and Lanos, P. (2003). Paleomagnetic directions and thermoluminescence dating from a bread oven-floor sequence in Lübeck (Germany): a record of 450 years of geomagnetic secular variation. *Journal of Geophysical Research*, **108**, 53–66.
URL http://www.agu.org/pubs/crossref/2003/2002JB001975.shtml

Silverman, B. W. (1984). Spline smoothing: the equivalent variable kernel method. *The Annals of Statistics*, **12**, 898–916.

Silverman, B. W. (1985). Some aspects of the spline smoothing approach to non-parametric regression curve fitting. *Journal of the Royal Statistical Society Series B*, **47**, 1–52.

Sternberg, R. (1989). Secular variation of the archaeomagnetic direction in the American Southwest, AD 750–1425. *Journal of Geophysical Research*, **94**, 527–546.

Stuiver, M. and Reimer, P. (1989). Histograms obtained from computerized radiocarbon age calibration. *Radiocarbon*, **31**, 817–823.

Tarantola, A. (1987). *Inverse problem theory: methods for data fitting and model parameter estimation*. Elsevier, Amsterdam.

Tarling, D. (1983). *Palaeomagnetism*. Chapman and Hall, London.

Tassi, P. (1992). *Méthodes statistiques*. Collection Economie et Statistiques avancées. Economica, Paris, second edn.

Thellier, E. (1938). Sur l'aimantation des terres cuites et ses applications géophysiques. *Annales de l'Institut de Physique du Globe de Paris*, **16**, 157–302.

Thellier, E. (1981). Sur la direction du champ magnétique terrestre en France durant les deux derniers millénaires. *Physics of the Earth and Planetary Interiors*, **24**, 89–132.

Thellier, E. and Thellier, O. (1959). Sur l'intensité du champ magnétique terrestre dans le passé historique et géologique. *Annales de Géophysique*, **15**, 285–376.

Tsunakawa, H. (1992). Bayesian approach to smoothing palaeomagnetic data using ABIC. *Geophysical Journal International*, **108**, 801–811.

Ulrych, T. J., Sacchi, M. D. and Woodbury, A. (1992). A Bayes tour of inversion: a tutorial. *Geophysics*, **66**, 55–69.

van der Plicht, J. (1993). The Groningen radiocarbon calibration program. *Radiocarbon*, **35**, 231–237.

van der Plicht, J. and Mook, W. G. (1989). Calibration of radiocarbon ages by computer. *Radiocarbon*, **31**, 805–816.

Wahba, G. (1990). *Spline models for observational data*. Pennsylvania, Society for Industrial and Applied Mathematics, Philadelphia, USA.

4

The Synchronization of Civilizations in the Eastern Mediterranean in the Second Millennium BC: Natural Science Dating Attempts

Otto Cichocki, Max Bichler, Gertrude Firneis, Walter Kutschera, Wolfgang Müller, and Peter Stadler

Summary. This chapter reports on work undertaken during the first three years of a ten-year project which aims to synchronize a range of relative and absolute dating evidence arising from archaeological records of the civilizations of the Eastern Mediterranean in the second millennium BC. At present the team is collecting chronological information from many different geographical locations. Some of the chronological methods are covered in more detail elsewhere in this volume (e.g. tephrochronology, Chapter 8 and radiocarbon dating with Bayesian models, Chapters 1 and 2) and other methods such as dendrochronology and astrochronology are explained in detail here. It is already clear that the different dating methods do not lead directly to a coherent chronological picture for the region. Consequently, one of the major issues that must be tackled by this project is the synchronisation of chronological evidence from different sources. This chapter outlines the nature of the evidence available and explains some of the techniques that the project team plans to use to link together the diverse dating evidence and thus develop their final chronological understanding.

4.1 Introduction: the SCIEM 2000 Project – Archaeological methods of creating and documenting chronology

The SCIEM 2000 project (a major research programme of the Austrian Academy of Sciences, the Institute of Egyptology at the University of Vienna and the Austrian Science Fund, FWF) started in 1999 and has recently completed the first three years of its planned ten years of existence (Bietak 2000). As the name indicates, the project's aim is the "Synchronization of Civilizations in the Eastern Mediterranean in the second millennium BC". In order to create a definite chronological framework for the history of the second millennium BC both archaeological and scientific methods are applied in 15 subprojects. The remarkable financial and personal effort is justified by the immense importance of the Middle and Late Bronze Ages for the rise of European culture and the history of the Ancient Near East, Egypt and Anatolia. The regrettable lack of knowledge concerning exact chronological data in this relevant period of time seems to be reason enough for a combined effort of all disciplines available. The first three years brought to light the gaps between different schools, sites and disciplines. It is the ultimate goal of our project to bridge those gaps by using both long-established and newly discovered links in material culture. Tell el Dab'a in the Eastern Nile Delta has been confirmed to be the site of ancient Auaris by Manfred Bietak and thus now offers a point of reference for our efforts.

The need for a new evaluation of the overall chronological situation became apparent especially in Egypt because High and Ultra High chronologies (Manning 1999) seemed to put a tremendous strain on the well-established Egyptian chronology based on the historically and epigraphically reconstructed kinglist. In spite of the fact that for most periods of Egyptian history names of kings together with the years of their reigns are known (Beckerath 1997; Kitchen 2000), significant gaps and uncertainties are still unsolved. Especially in the Second Intermediate Period the consistency of the historical Egyptian record shows serious lapses. There is much discussion on co-regencies and other possibilities that tend to weaken the relevance of the Egyptian list for absolute dating in this part of the second millennium (Ryholt 1997). The chronology of the Middle Kingdom on the other hand, is generally assumed to be accurate. The absolute chronological dates are gained by calculating the known lengths of regencies and taking the reported and recalculated Sothis dates as fixed points. The traditional one-sided picture of the Egyptian part of the link producing the absolute dates for Aegean and other neighbouring sequences seemed to change when the Santorini eruption was dated with scientific methods. Unfortunately this new, very early date (17th century BC) seemed to make the sequences drift apart. It appears to be quite impossible to squeeze an additional 150 years out of the traditional sequence of time based on the regencies of Egyptian kings. Scholars who were used to chronological discrepancies of 20 to 30 years suddenly saw themselves confronted with a completely new, utterly irritating situation.

There are several important reasons for the prominent position of Tell el Dab'a in the project. First of all it is a node in the network of Bronze Age interrelations, ethnically, culturally and with respect to the material culture of the Eastern Mediterranean. This fact has been established by long, continuous excavations that have tried to meet the highest standards both of excavation technique and documentation from the beginning (Bietak 1976). A population of Levantine origin dwelt in the Eastern Delta together with Egyptians for hundreds of years and produced a material culture that shows both Egyptian and Levantine traits. Recent finds of Minoan wall paintings and weapons of Aegean origin gave further evidence of the multicultural nature of the site.

Hopes for the linking of the main sites of the Eastern Mediterranean are pinned on the Natural Sciences. The existing chronology had seemed sufficient to most Egyptologists. Therefore few samples for scientific investigation had been gathered on excavations and in museums in Egypt. Sites outside Egypt developed independent relative chronologies being dated in terms of absolute chronology by means of sometimes sporadic Egyptian finds. For example the chronological sequences for Crete and Cyprus have been established by means of Egyptaica occasionally bearing names of kings. The Natural Sciences offer new opportunities. Besides physical processes that are nearly the same everywhere (e.g. ^{14}C), both Egypt and the Eastern Mediterranean should have been affected by the Santorini eruption at exactly the same time (Manning 1999). This effect may serve as a time line.

Numerous conferences, meetings and workshops have shown that an accumulation of as many perspectives as possible is most likely to lead to a framework for chronologically pin-pointing the traces of human activity that the different researchers find on their excavations.

Those traces of human activity materialize as artefacts, weapons, pottery, other parts of material culture and structures: architecture, tombs, etc. The context of these artefacts, their position and connection to other structures form the "containers" that are called stratigraphic units and are usually combined with a local relative chronology by means of certain signifiers being isolated and treated as "index fossils". These significant parts of material culture can be found on other sites, sometimes far away from their place of first appearance.

The "first appearances" of special items have to be recorded and compared. New shapes of vessels for instance may well be hints towards a change in material culture. Synchronization of such changes, preferably backed by scientific data, might be of great importance for the construction of a new supra-regional chronology. A branch of the SCIEM 2000 project, called "Stratigraphie Comparée" collects data, preferably on site, and tries to produce a synopsis of confirmed synchronisms by using Tell el Dab'a and Ashkelon as key sites. The project has succeeded in gaining the co-operation of eminent institutions and scholars working in the field, including the University of Rome (Paolo Matthiae), the Deutsches Archäolgische Institut Istanbul (Felix Blocher), the French Academy and Institut Français d'Archéologie du Proche-Orient, Lebanon (Jean Paul Thalmann).

Amongst the pottery the Middle to Late Cypriote pottery is especially significant for the construction of timelines (Maguire 1992, 1995) and is therefore investigated by a special branch of our project.

Besides the external factors of material culture, i.e. the circumstances of disposal, transport, use, etc. (in short, the context of pottery and other finds), the factors that are inherent in groups of material culture are also investigated by comparing features and measurements. Arranging the types according to their degree of similarity produces a sequence that can be correlated to a gradual chronological development. Quantitative methods like seriation are applied to groups of pottery, like cups, or to weapons (Bietak 1985).

The traditional way of arranging objects in sequences is the typological method. The contexts that are required for the construction of typologies in the Montelian sense have been found in Tell el Dab'a in the form of numerous tombs, some of them undisturbed. Recent work tries to focus on the problematic nature of this way of creating chronological sequences. The linking of highly "ideological" deposition in the funerary context to settlement contexts may produce an incorrect picture of chronological development.

The Data Management Branch of the SCIEM 2000 project constitutes a node within the project where scientific and archaeological efforts are fused with the help of a specially developed software package. Flexible data management and the necessity of a certain degree of standardization are important features of modern relational database design and help to enforce terminological and other conventions while providing abundant possibilities for output and analysis. The more general aim of constructing databases for input of all relevant data has been approached during the first part of the project and will now be expanded further to new projects, like the so-called "Handbook of Middle Kingdom Pottery". This is intended to be a guidebook where any pottery shapes found in the Eastern Mediterranean can be compared with the corpus of Middle Kingdom pottery found in Egypt. This corpus will be as complete as possible. Both a traditional publication in two volumes and an interactive database accessible via the Internet are planned. The database is based on traditional typological labels and newly developed shape-recognition software that allows a user to access the database, and to compare a shape with the corpus even in ignorance of the typonyms of Egyptian Archaeology, with the degree of similarity evaluated by means of a seriation package. The archaeologist is thus able to find the links in material culture to verify whether a given vessel is of Egyptian origin. If it turns out to match then information as to where and how the vessel clusters within the corpus of Egyptian pottery is promptly available. Databases on the pottery of other epochs of Egyptian history are planned for the future.

For more information on projects, collaborators, conferences, publications see our homepage at http://www.nhm-wien.ac.at/sciem2000/.

Archaeological evidence from nine archaeological projects is currently being investigated and compared. Particular emphasis is being placed on the synchronization of the Egyptian chronology of the Middle Kingdom and Second Intermediate Period with the period of transition from Early Minoan/Middle Minoan to Late Minoan strata in the Aegean. The aims and possible problems are discussed at length in the proceedings of an international symposium at Haindorf Castle in 1998 (Bietak 2000).

To produce absolute data, independent from the relative results of archaeology, four science projects working with different materials and approaches are involved. They are the focus of this chapter.

4.2 The Thera Ashes Project

The general idea of this project is to fix the "Minoan" eruption of the Santorini volcano during the 2nd millennium BC within the relative chronologies of the regions of the Eastern Mediterranean by finding the eruption products of this event within well-defined stratigraphies of archaeological sites. Although the date of the eruption is still the subject of intense debates, the eruption products, wherever they are found, can be used as tangible evidence for con- temporaneity or at least post-eruption dating of the strata. During this strong explosive event a large volume of magma was erupted in a short time span of not more than a few days and the eruption products were distributed over a large area. The majority of the erupted material consists of chemically rather homogeneous pumice tephra and pumiceous flow deposits, the "Minoan tuff" or "Oberer Bimsstein" (Bo, upper pumice; Vitaliano et al. 1978, 1990; Fran- caviglia and DiSabatino 1990; Sparks and Wilson 1990; Druitt et al. 1999). The volume estimations range from 16 to 35 km^3 of dense rock equivalent (Sig- urdsson et al. 1990; Pyle 1990). The impact on the contemporary civilizations is evident and alluvial pumice as well as direct fallout from the eruption cloud is reported from several sites on Greek islands and Asia Minor (Marinatos 1932; Zeist et al. 1975; Watkins et al. 1978; Keller 1980; Pichler and Schiering 1980; Vinci 1985; Francaviglia 1986, 1990; Sullivan 1988; Marketou 1990; Soles and Davaras 1990; Warren and Puchelt 1990; Guichard et al. 1993; Eastwood et al. 1999). The fallout formed a synchronous layer of volcanic ash (tephra with a grain size of less than 2 mm) that can be used directly as a datum line. Assuming a reliable identification, it can be used for chronology wherever it is found in primary deposits (Einarsson 1986). Furthermore, pumice is a very useful abrasive and has been collected and traded since prehistoric times, as reported by Pliny in his *Natural History* book XXXVI (Faure 1971). The ap- plicability for chronological purposes has been checked in earlier studies by demonstrating that the "Minoan" pumice is sufficiently homogeneous and can be distinguished from the numerous other Aegean pumice sources by its trace element distribution pattern ("chemical fingerprinting"; Bichler et al. 1997; Peltz et al. 1999; Schmid et al. 2000). The identification of distant tephra layers and the possible sources of errors are discussed in Schmid et al. (2000) and Saminger et al. (2000).

4.2.1 Pumice Results

Ninety-six samples of archaeologically stratified pumice found in excavations in Egypt, Palestine and Israel were investigated and identified. One sample from Tell el Dab'a could not be related to an Aegean volcanic source. Their geographical locations are shown in Figure 4.1, a sample overview is given in Table 4.1.

The sample preparation is described in detail in Peltz et al. (1999). The analytical results allow a clear identification of the volcanic sources (Bichler

Fig. 4.1. Geographical locations of pumice sources in the Aegean Sea and archaeological excavation sites with finds of pumice.

Table 4.1. Identified pumice samples. Period abbreviations are MB: Middle Bronze Age; LB: Late Bronze Age; LC: Late Cypriote.

Excavation site	Period	Number	Provenance
Tell el Dab'a	MB/LB	32	1 Nisyros, 29 Santorini, 2 Kos
Tell el Herr	Roman	1	Santorini
Tell el Hebwa	LB	10	all Santorini
Ashkelon	MB	3	1 Giali, 2 Nisyros
Tell el Ajjul	MB/LB	48	46 Santorini, 2 Nisyros
Maroni	LC IIC	1	Santorini
Lachish	13th cent. BC	1	Nisyros

et al. 2002). To graphically demonstrate the identification, a normalization procedure is carried out; the concentration of an element is divided by the mean concentration of that element in bulk Minoan pumice. Thus the graphs show only the deviations from the composition of the Minoan pumice.

First, investigations on well-defined volcanic strata of several Aegean volcanoes were carried out to clearly define their individual chemical compositions. Figures 4.2 to 4.5 show the natural variation ranges of these Aegean pumice sources and the agreement with some typical samples from excavations. Our results demonstrate that the pumice samples found in archaeological excavations only show minor changes in composition due to leaching effects, although glass hydration effects and strongly weathered rock surfaces were observed, especially in the water-saturated sedimentary environment of Tell el Dab'a. The samples identified as related to the Kos Plateau Tuff eruption show a significant depletion of some elements (Sc, Co and Sb). This is due to the special mineralogical properties of this material and does not interfere with the identification (see Peltz et al. 1999). From the chronological point of view, the results obtained for the excavated samples agree with the present dating range of the Minoan eruption between 1650 and 1450 BC.

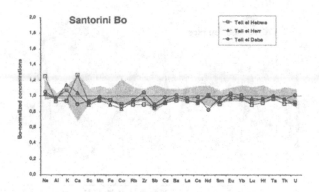

Fig. 4.2. Normalized element abundances of pumice samples from Egypt identified as "Minoan" pumice from the Santorini Bo-layer. The shaded area shows the natural range of variation of elements in Bo-pumice. Normalization factors and natural variations from Peltz et al. (1999).

4.2.2 Volcanic Ash Results

Tephra particles have been sought in 114 stratified soil samples from Miletos, Ebla, Megiddo and Cyprus. Additionally, drill cores from Delphinos (Crete,

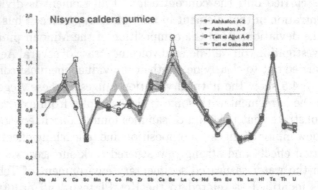

Fig. 4.3. Normalized element abundances of selected elements found in pumice samples from Egypt and Palestine identified as pumice from Nisyros. The shaded area shows the natural range of variation of elements in Nisyros pumice. Normalization factors and natural variations from Peltz et al. (1999).

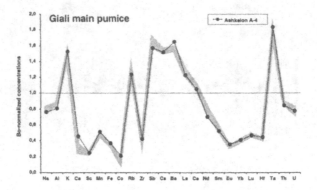

Fig. 4.4. Normalized element abundances of selected elements found in a pumice sample from Ashkelon identified as pumice from Giali. The shaded area shows the natural range of variation of elements in Giali pumice. Normalization factors and natural variations from Peltz et al. (1999).

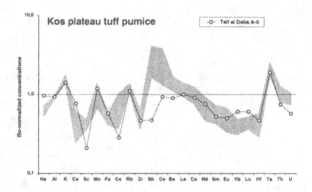

Fig. 4.5. Normalized element abundances of selected elements found in a pumice sample from Egypt identified as pumice from Kos. The shaded area shows the natural range of variation of elements in Kos pumice. Normalization factors and natural variations from Peltz et al. (1999).

Greece), Ova Gölü (south-western Anatolia) Gölbasi and Amuq (Orontes plain, Turkey) have been sampled. The geographical locations are shown in Figure 4.6. The preparation included a separation procedure (decarbonatization, sieving fraction 0.01 to 0.25 mm) to enrich tephra particles prior to polarization microscopy. Only five soil samples, all from Miletos, and one drill core (Delphinos) were found to contain volcanic particles. From these samples a satisfying pure tephra fraction (>95%) has been separated to enable the analytical identification. The results certify a Santorini "Minoan" origin (see Figures 4.7 and 4.8). The enrichment of chromium and antimony in the Milesian samples is caused by local soil contamination.

4.2.3 Discussion of Thera Ashes Results

On the basis of the results obtained we draw the following conclusions: the area, where direct fallout from the eruption plume could be expected in detectable quantities in prehistorically inhabited regions, is restricted to Aegean islands and Asia Minor. It can be assumed that low-altitude tropospheric currents led to a "tephra fan", which is in accordance with earlier estimations based on deep-sea drill cores (Richardson and Ninkovich 1976; Pyle 1990). Fine particles reaching higher atmospheric levels obviously followed a different, ENE-oriented distribution trend (Figure 4.9). This agrees perfectly with the reported tephra findings in Anatolia and the Black Sea (Guichard et al. 1993; Eastwood et al. 1999; Pearce et al. 1999). Previous studies (Stanley and Sheng 1986) that report tephra particles from the "Minoan eruption" even

92   Otto Cichocki *et al.*

Fig. 4.6. Geographical locations of soil samples and drill cores.

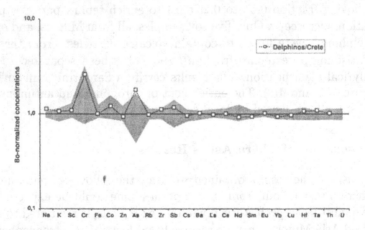

Fig. 4.7. Normalized element abundances of volcanic glass particles separated from the Delphinos (Crete) drill core identified as "Minoan" tephra. The shaded area shows the natural range of variation of elements in Bo-pumice. Normalization factors and natural variations from Peltz et al. (1999).

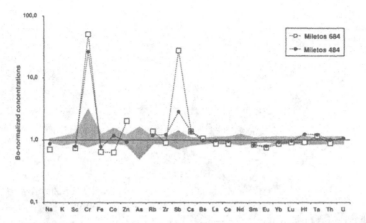

Fig. 4.8. Normalized element abundances of volcanic glass particles separated from soil samples from Miletos, identified as "Minoan" tephra. The shaded area shows the natural range of variation of elements in Bo-pumice. Normalization factors and natural variations from Peltz et al. (1999).

in Nile delta sediments are not necessarily erroneous, but might have identified particles produced by fragmentation of alluvial pumice deposits or even by human use of pumice as an abrasive. The numerous findings of "Minoan" pumice in Tell el Dab'a prove the extensive use of this pyroclastic material in more or less contemporary Egypt. Consequently, the search for pumice in settlements being excavated has been intensified and yielded a large number of samples in cooperation with the archaeologists responsible. Surveying the first appearances of "Minoan" pumice in well-stratified sites will enlarge the region where the products of this eruption can be used to establish a datum line across the Eastern Mediterranean and will contribute significantly to the determination of the absolute chronology of the eruption.

4.3 The Astrochronology Project

4.3.1 Introduction

According to the astronomer Sosigenes the ancient Egyptians had been developing their own calendric system since about 3000 BC and started off with a lunar calendar. Egypt is unique in early civilizations in being especially dependent on one event: the flooding of the Nile. The moon consistently failed to predict this but the stars were very serviceable. The Egyptian priests soon noticed that the heliacal rising of Sirius (first day visible just before sunrise) always preceded the flood by a few days (Library 1983; O'Neil 1986; Paul 1993; North 1994). They eventually had a system of 36 stars to mark out the year and ultimately had three different calendars working concurrently

Fig. 4.9. Proposed regional distribution of the Minoan tephra deposit.

for over 2000 years: a stellar calendar for agriculture, a solar year of 365 days (12 months × 30 + 5 extra) and a quasi-lunar calendar for festivals.

In the last decades BC the calendar had become so confused that Julius Caesar invited the astronomer Sosigenes from Alexandria to renew the basic system for both the Roman Empire and Egypt. Sosigenes replaced the lunar system with a tropical year of 365.25 days (the *solar*, or *tropical*, year is the time interval between successive passages of the sun through the vernal equinox, which is the time of the year when the lengths of the day and the night are equal). Further, to correct the accumulation of previous errors, a total of 85 intercalary days had to be added to 46 BC, meaning that 1 January 45 BC occurred in what would have been the middle of March. To prevent the problem from recurring, Sosigenes suggested that an extra day be added to every fourth February. The adoption of such reformatory measures resulted in the establishment of the Julian calendar, which was used for approximately the next 1600 years (Richmond 1956; O'Neil 1986; North 1994).

As some political events in the Middle and New Kingdom are mentioned in connection with stellar or lunar constellations, the aim of this project is to recalculate these constellations to achieve more precise astronomical dates for these events.

4.3.2 Calculation of Stellar and Lunar Constellations

The present investigation creates a database for heliacal risings of Sirius calculated for the latitudes of Memphis and Thebes during the years 2000 BC until AD 1. The tables are calculated with the *java-script* program *HELIAC* which offers sufficient accuracy to give the correct day of this stellar phenomenon (http://geocities.com/CapeCanaveral/Launchpad/4633/heliacJAVA.html).

Depending on the transparency of the atmosphere, heliacal risings can occur when Sirius is 3° to 5° above the horizon. The recalculated results for a height of 3° are in perfect agreement with other recent recalculations (Schaefer 1997), where atmospheric and aerosol components were incorporated as well as an extinction-coefficient.

In order to archive a better temporal correlation with the monthly festival texts of the Illahun tablets, a database for the dates of the last lunar crescent visibility and the first lunar crescent visibility for the years 1861 to 1811 BC are given at the latitude of Thebes. No corresponding tables are given for Memphis, the other astronomical centre of ancient Egypt, as the lunar heights at these special phases differ only by a mean value of 0.3° and thus do not yield any different monthly dates. The program *URANIASTAR 1.1* (available at http://members.eunet.at/vollmann/) was found to yield the correct dates when compared with the most recent lunar theory by Chapront-Touzé and Chapront (1991). Further special lunar quarter phases can be found in the Five Millennia Catalog of Phases of the Moon −1999 to +3000 calculated by F. Espenak (in Chapront-Touzé and Chapront 1991, and at http://sunearth.gsfc.nasa.gov/eclipse/phase/phasecat.html).

4.3.3 Discussion of Astrochronology Results

Three major uncertainties influencing the interpretation and usage of the reconstructed stellar and lunar constellations still exist:

- When was the actual beginning of the Old Egyptian month? Was it definitely the last lunar crescent?
- Where were the astronomical observers based? Were they working at Memphis or Thebes or somewhere else?
- As the heliacal rising of Sirius was reported only every fourth year, when was the beginning of this cycle? Exactly when did the five extra days per year fall?
- Was the heliacal rising of Sirius defined to happen with an angle of 3°, 4° or 5° above the horizon?

It is not yet possible to find any detailed answers to these questions. Thus the accumulation of all these uncertainties makes it impossible to achieve a resolution better than the time difference between the "High Chronology" and the "Low Chronology". Therefore the results of the project will only be helpful to give more accurate absolute astronomical dates if based on improved historical knowledge of details (definitions, methods, and locations) of Ancient Egyptian astronomy.

4.4 The Dendrochronology Project

4.4.1 Introduction

The main aim of this project is to carry out dendrochronological investigation of wooden artefacts (building construction parts, coffins, picture tables, objects of art, charcoal) and to establish floating chronologies for the second millennium BC. The aim is to link the objects of the relevant time period to a relative chronology of certain historical events. These efforts are being continued in order to close gaps between existing floating chronologies.

By autumn 2002 we had started collecting data and samples from standing trees, historic buildings and archaeological finds (charcoal, ship's masts, etc.) in an attempt to set up an absolute standard starting in the present and reaching back to the second millennium BC. It would then be possible to establish absolute dates for other wooden objects via the age of the last ring (if the waney edge, which is the last ring grown before the tree's death, is preserved).

4.4.2 Selection of Wood Species for Dendrochronological Investigations

The selection of wood species for investigatation depends on their occurrence in the archaeological objects available for investigation and on their suitability for dendrochronology.

Table 4.2. Compilation of results of wood species analysis of Old Egyptian objects from the Louvre (Nibbi 1981), Munich and other collections (Grosser et al. 1992) and the British Museum (Davies 1995).

Wood species	Louvre	Munich and other	British Museum
Ficus sycomorus	32	138	230
Tamarix sp.	> 100	106	158
Acacia sp.	35	42	53
Cedrus sp.	5(7)	44	88
Juniperus sp.	2	10	16
Pinus sp.	9	3	37
Cupressus sp.	–	2	7

The results shown in Table 4.2 give a rough idea of the spectrum of wood species used in ancient Egypt for various purposes but do not necessarily reflect the spectrum of samples available for the project.

Cedrus libani is perhaps the most interesting of the species cited below. This species grows in Mediterranean mountain climates and has distinct ring borders. Its life span is said to be as long as 500 years. It has been imported to Egypt since the time of the Old Kingdom. Due to over-deforestation only a few small relict areas are still alive in Lebanon, with larger ones in Turkey (Taurus, Antitaurus) and Syria. The subspecies *Cedrus libani atlantica* grows in the Atlas Mountains of Algeria and Morocco, whilst *Cedrus libani brevifolia* grows in a very limited area in the mountains of Cyprus. These three subspecies cannot be separated by anatomical features of their wood. Hence, because of different interpretations of Egyptian texts mentioning imports of wood from certain countries, a heated debate continues as to whether cedar wood found in Egypt had its origin in Lebanon or somewhere else (Nibbi 1981, 1987, 1994, 1996; Meiggs 1982; Grosser et al. 1992; Davies 1995). Another debate deals with the correct translation of different hieroglyphs which characterize different wood species (Loret 1916; Meiggs 1982).

The degree of deforestation can be estimated by the occurrence of rock carvings marking so-called "Hadrian's forests" (Mikesell 1969). These forests were marked and protected as a resource for building ships for the Roman navy. The area of cedar forests was much larger even in Roman times than it is today.

Substantial investigations on important archaeological sites and the occurrence of cedar wood were carried out by Liphschitz (1986), Bikai (1991), Lev-Yadun (1992), Liphschitz and Biger (1992) Kuniholm (1996), and Newton (1996). Unfortunately, in the literature the term "cedar" was used not only for *Cedrus* sp., but also for wood of *Juniperus virginiana* (Lucas and Harris 1962), as well as a modern commercial name for wood of the Cupressaceae *Calocedrus decurrens*, *Thuja plicata* and *Thuja occidentalis* (Wagenführ and Schreiber 1989).

4.4.3 Objects to be Measured

For a floating chronology (2nd millennium BC): these will mainly be objects of art stored in museum collections (coffins, painted panels, statues etc.), but also new finds in recent excavations (e.g. charcoal).

For absolute standard (from today back to the 2nd millennium BC): samples will mainly be taken from standing trees, construction materials (roof constructions, ceilings, wall supports etc.) of standing buildings, but also from objects of art.

4.4.4 Sampling and Data Acquisition

In most cases it is not recommended to mix samples of different wood species for the construction of a reference chronology. To avoid these problems wood species analysis (identification of anatomical structures of a very small sample with the help of a light microscope) of all objects available for investigation is the first step.

In order to measure the thickness of all the rings in a particular piece of wood, it is necessary to get a view of its cross-section. Wooden parts of the architecture of a building often allow either a disc to be cut or a radial core to be drilled. After smoothing the surface, the ring borders become visible. For our project in some cases a 14 mm coring tool may be used, as cores can be polished as finely as necessary and doubtful measurements may be re-examined later.

If the surface (square or longitudinal) of a wooden object is not covered with paint and has been well smoothed by the ancient manufacturer, after some cleaning the rings can often be measured directly on the object. For objects coated with paint or other surface covering, equipment for drilling a 5 mm diameter hole to allow endoscopic measurement of the ring widths is under construction.

At least two radii of a sample are measured. The mean is calculated to compensate for biological diversity (e.g. elliptic ring shape, single growth deviations). The result is a list of mean growth-ring thickness for each year contained in the sample.

A sample has to show at least 50 rings, otherwise the statistical methods for dating will fail to work. It is necessary to collect and measure as many promising samples as possible from one complex (a "time-unit") to calculate a reliable mean value list for further analysis.

4.4.5 Measuring the "Samples"

In the beginning we had hoped to use the techniques usually used in dendrochronology to collect data (to cut discs or drill cores), but sampling of any museum objects turned out to be impossible. We might have been permitted to drill the occasional core, but without lots of data a statistical method like dendrochronology does not work. Therefore we were forced to develop non-destructive methods to collect a sufficient amount of accurate data. Different techniques were tested to measure the ring-width on plain and unpainted longitudinal and square surfaces. Two of them are already working properly.

For plain surfaces a flat-bed scanner with 2400 dpi resolution was redesigned to work in all positions. It is placed on a monopod or tripod and the scanning direction is orientated parallel to the ring borders. A strip of approximately 4 cm is scanned and stored on the hard-disc of a laptop. It can be studied immediately for sharpness and visibility of all ringborders. The measurements are made later with the help of the picture analysis program OSM3 especially designed by Bernhard Knibbe (for details see http://www.sciem.com/main.html). For each scan a record of collection, object, object number and piece of wood is documented in a database.

For bent surfaces, or situations where special illumination or flexibility of positioning are required, a machine was constructed using a video camera with a magnifying lens moved using a sledge driven by a step motor. This device allows us to sight a ring border on the magnified picture on the laptop and then to move the sledge to the next border. The distance the sledge has been moved is measured to an accuracy of 0.01 mm with the help of the measurement program PAST32 (designed by Bernhard Knibbe, details again at http://www.sciem.com/main.html) and is equivalent to the thickness of the measured ring. Simultaneously a picture of each measurement position can be saved to the hard-disc of the laptop to be studied later in dubious cases. Although positioning of the equipment and measurements take much longer and have to be done on the spot, many objects not accessible by scanner (for example those with curved surfaces) can be investigated.

4.4.6 Dating Procedure

Two samples grown at the same time in the same climatic conditions will show a statistically significant correlation of their distribution of thick and thin rings (their mean value lists). Such a pattern is unique to a certain time period, a certain climatic region and, unfortunately in many cases, only within one genus (or species) of tree. Cross-dating (statistical comparison of two mean value lists) seeks to synchronize different samples and may find the relative time span between them.

If two samples do not completely overlap in time, their combined matched pattern will be longer than one of the single samples. Many successfully cross-dated overlapping samples from different periods – the youngest one being of known age (e.g. a recently cut tree) – allow a tree ring standard (a dated chronology) to be defined which is valid for a certain wood species and a limited growth area. If a dated sample is missing, the result is a so-called floating chronology.

Dating a sample means comparing all overlapping positions of a sample with a standard and finding the best matching position. In many cases there is more than one position possible from a single statistical analysis. To find the correct position "pointer years" and various statistical tests are used.

Gleichläufigkeit

This value represents the percentage of slope equivalence of a given sample and reference within the overlapping parts of these two records. The Gleichläufigkeit (which can be translated as "synchronicity value", but according to Kaennel and Schweingruber 1995, the expression "Gleichläufigkeit" is also used in English due to lack of proper translation) is calculated in the following manner. Both reference and sample values are "digitized" in one-year steps. Possible values are −1 for decreasing slope, 0 for no change and +1 for years with increasing values. The digitized values of the overlapping parts of reference and sample are compared and one-year intervals with similar slopes are counted. The ratio of the resulting value to the number of overlapping years gives the Gleichläufigkeit value (0 to 100%).

Pointer Year Calculation

When setting up the standard by cross-dating all its samples, certain years form pronounced peaks in the graph. If, for example, more than 75% of all rings belonging to one year have the same trend (increasing or decreasing growth compared with the ring before), this year with especially strong influence on growth is called a "pointer year" and is specially marked. In dubious cases, in a second match, only these pointer years are compared with the corresponding rings of the sample. The correct position will then show a significantly higher correlation of these special years than the other positions in question. If the reference record is a chronology with stored density data, the synchronicity value of pointer years represents only the pointer years, which should exceed the overall synchronicity value.

t-Test

An adapted Student's *t*-test is performed with the sample and reference records as two data sets. The results measure the degree of similarity of two data sets in a certain position to each other.

Skeleton Plot

The Skeleton Plot view is a traditional approach to check for the integrity and fitting of two records. Usually the smallest ring of the sample will be drawn as a long bar, small rings with small bars, rings with average width and wide ones are not drawn. This method helps to solve the problem of missing rings, but has no formal statistical approach associated with it.

As these different approaches have been developed worldwide and step-by-step and are used differently (for example the skeleton plot is not used in Europe), documentation in literature is widely scattered. A good example for a modern approach of American methods of synchronization is the study of Engle (2000).

Measuring Problems and Dating Results

In years with bad growth conditions it happens with many species that one or more rings are completely missing or are not present all the way around the circumference of a stem or branch. If this missing ring is not detected the sample will not match well with the standard. Of course, this is even more misleading if it occurs when the standard is being set up in the first place. Detection is possible by comparing many samples (usually the same ring is not missing in all of them). The best possibility for detecting incomplete rings is given on cut discs by comparing different radii.

In years with a temporary period of low temperature or drought during the growing season a "false ring" may occur. But this "latewood formation" within the ring has a much smoother outer border than the real latewood formed at the end of the growth season. Hence, it can be detected under the microscope.

Trees grow outwards from their centres one annual ring at a time. To avoid confusion, the age of a wood sample is usually defined as the absolute age of the outermost ring preserved in the sample. If this ring is the last one grown before death of the tree (the so-called waney edge, in some cases preserved with bark), it is possible to date the felling year and also to identify the felling season.

Problems can arise from samples of wood in secondary use or an undetected repair, for they may give a false higher or lower age respectively for the complex within which they were found (for this and other reasons mentioned above, it is of advantage to sample as many promising specimens as possible).

4.4.7 Discussion and Future Work for Dendrochronology

In the first three years of "Project 7 / Dendrochronology", data have been acquired at Kunsthistorisches Museum, Vienna (1 object), Egyptian Museum, Cairo (14 objects), Metropolitain Museum of Art, New York (13 objects), Museum of Fine Arts, Boston (49 objects) and British Museum (16 objects.). Most objects consist of several pieces of wood, each usable piece was measured two or three times.

The data already acquired have been synchronized and combined to create floating chronologies. At present they cover 607 years of the 2nd millennium BC and several shorter sequences; another longer chronology for the New Kingdom is under construction.

Besides the museum objects two other sources are now under investigation. In autumn 2002 we collected charcoal samples from excavations at Arqa (Lebanon) and Qatna (Syria), which are positioned by archaeologists in the 2nd millennium BC. As parts of the collection have arrived in Vienna in spring 2003 the Qatna samples show extremely narrow rings, which requires us to improve our measurement equipment. As both collections contain wood definitely grown in the Lebanese mountains, the comparison of this data set with the Egyptian Museum data will prove whether the Egyptian cedar imports really come from this region.

The museum data are producing doubts in this respect. Most of the objects consisting of several pieces produce more than one floating chronology. This may be the result of mixing wood of different ages or of contemporary wood grown under different ecological conditions. Studying the Lebanese and Syrian samples may help to understand from a dendrochronological point of view the uniformity or complexity of contemporary wood grown in this region.

Another approach is to collect samples and data to construct a cedar standard for the Lebanese area for producing absolute dates. The sources will be mainly standing or fallen trees from relict areas, wooden parts of buildings (lintels, doors, beams, roof elements), furniture and finds from excavations. To bridge the time span between the Middle Ages and the Roman period will require hard work combined with good luck.

4.5 ^{14}C Dating Project

4.5.1 Introduction

The aim of this project is to contribute to the synchronization of Eastern Mediterranean cultures by using ^{14}C dating with accelerator mass spectrometry (AMS), in an interdisciplinary initiative between archaeologists and nuclear physicists. An improved absolute chronology based on precise ^{14}C dating will lead to a better understanding of the interactions between the cultures of the Eastern Mediterranean in the second millennium BC. The ^{14}C dating is performed at the Vienna Environmental Research Accelerator, a centre for AMS at the Institute for Isotope Research and Nuclear Physics of the University of Vienna, which came into operation in 1996 (Wild et al. 1998). With the help of this facility it is now possible to compete with the best ^{14}C AMS facilities in the world, obtaining radiocarbon ages with a precision between ± 30 and ± 40 years.

4.5.2 State of Research

In order to measure radiocarbon ages it is necessary to find the amount of radiocarbon in a sample. This can be achieved either by measuring the radioactivity of the sample (the conventional beta-counting method) or by directly counting the radiocarbon atoms using a method called accelerator mass spectrometry (AMS). The main advantage of AMS over the conventional beta-counting method is the much greater sensitivity of the measurement. In AMS the radiocarbon atoms are directly detected instead of waiting for them to decay. The physical sample sizes required are typically 1000 times smaller, allowing a much greater choice of samples and enabling very selective chemical pre-treatment. However, handling small amounts of sample material increases the danger of contamination. Since minute additions of non-genuine carbon can lead to incorrect results, the entire sample preparation has to be performed with utmost care, which is time consuming.

Among a variety of selection criteria for samples, two are particularly important. First, the samples should have a good archaeological context, which means that their position in the context should be known and the stratigraphy should be clear. Second, these strata or contexts should also contain well-defined remains, which can be dated by archaeological methods. Only in special cases should contexts without a good archaeological dating be dated with ^{14}C (e.g. graves without grave-goods).

Careful sampling and pre-treatment are very important stages in the ^{14}C dating process, particularly for archaeological samples where there is frequently contamination from the soil.

An important issue with ^{14}C dating is calibration. We use the *OxCal* program from the University of Oxford for this purpose (Bronk Ramsey 1995, 1998, 2000, 2001; Bronk Ramsey et al. 2001, and at http://www.rlaha.ox.ac.uk/orau/06_ind.htm).

In contrast to the many ^{14}C dates obtained in the past, without comparable standards of sampling and scattered in many archaeological publications, the aim of this project is to measure a large number of systematically selected samples from well-defined contexts or layers from the whole Eastern Mediterranean from Early Minoan/Middle Minoan to Late Minoan times. Some 48 samples have so far been measured; other samples from excavations in Milet (collected by W.D. Niemeyer) and from excavations in Egypt and Israel/Palestine are in preparation and will be analysed during this year.

4.5.3 Methods

The samples will be selected by the collaborators on the project with respect to the stratigraphy of the respective sites (in the light of the guidance given by Bayliss and Bronk Ramsey in Chapter 2).

A new methodology is being developed to further refine high-precision dating with ^{14}C wiggle-matching by the inclusion of external information (tree-rings, relative chronology by seriation or stratigraphy). For a single date the uncertainty in the calibrated date is, in general, dominated by the "wiggliness" in the calibration curve, and is related to a lesser extent to the precision of the ^{14}C measurement. Although the wiggles are really an obstacle for obtaining a precise absolute date with only a single ^{14}C date, one can take advantage of these wiggles if a series of radiocarbon dates is available. If such a series is combined with external information using Bayesian statistics, a considerable improvement of absolute dating can be reached. The method is also called wiggle matching. It is implemented in *OxCal*. So far, we envision four main applications in wiggle matching:

1. Floating dendrochronologies with a series of ^{14}C samples can be wiggle-matched by using a known tree-ring distance (in years) between the different samples (Christen and Litton 1995).
2. Seriation results are relative chronologies obtained by different methods of combination statistics. If one takes a series of ^{14}C samples one can see if the sequence from seriation correlates with the measured years BP. If there

exists such a correlation one can determine the slope of the correlation in SD/year BP. With the help of this information a "time span" can be calculated between each of the samples, including a sigma uncertainty for the time distance. This information can be used in wiggle matching with *OxCal* procedure V_Sequence (Christen and Litton 1995).

3. In a similar way stratigraphies may be ^{14}C dated. One samples the different layers in a stratigraphy for ^{14}C and takes into account also depth information. If the sedimentation process was continuous one may get a correlation between depth and years BP (Christen et al. 1995). In most cases this may not be possible. If it is, however, one can calculate a sedimentation rate in cm/year. One may then proceed in the same way as described above.

4. If the stratigraphy provides only information about which layer is older and which is younger, and if the sedimentation process was not continuous, there is also a possibility of using Bayesian statistics to shorten the time span of a calibrated sample in a series of ^{14}C dates (Buck et al. 1992).

The precision of the methods and reliability of results decreases from 1 to 4. We expect that these procedures will allow us to date a series of ^{14}C samples to a precision of ±15 years at the 68% confidence level after calibration.

In *OxCal* procedures 2 and 3 can only be obtained by using external programmes at the moment, thus the realization is complicated. Therefore we will suggest improvements of *OxCal* to Christopher Bronk Ramsey at Oxford University, or, alternatively implement these additions ourselves. The goal is to better combine the four different wiggle matching applications with archaeological methods.

4.6 Process of Combination of the Results of Different Projects

As the discussion of several controversial theories about Egyptian chronology has become quite heated, the main focus of all our projects has to be to produce data that are as independent as possible from one or other of the hypotheses. A compilation at this stage of producing data might unduly influence the interpretation, so this procedure is planned in a later stage of the projects.

To solve the discrepancies of old ^{14}C data with known archaeological contexts new approaches will be adopted (Buck et al. 1991, 1994).

Of course each project tries to support the attempts, and to reconsider the results of the others. For example a pumice layer in the complex stratigraphy of the Miletos excavation investigated by the Thera Ashes project was shown to be a Thera product. Now a sequence of charcoal samples from the same archaeological profile is ready for ^{14}C measurement to fix this pumice layer in time.

The same will happen with the charcoal from Arqa and Qatna. After preparation and dendrochronological measurement, the floating chronologies will be placed in time by ^{14}C to have a rough idea of their relative position and to check the correct synchronization of the single measurement sequences. After combination of many overlapping chronologies, floating in different time spans, the final aim is to develop a Cedar standard to arrive at absolute and precise-to-the-year data.

The comparison and synchronization of archaeological and scientific results will need to be even more sophisticated. We are aware that this final compilation attempt, and not only single project outputs, has to be the main result of all efforts. A successful (i.e. widely accepted) compilation might either support one or other of the existing relative chronologies or create a new system.

4.7 Contacts

The project leaders are:

- Thera Ashes: Prof. Dr. Max Bichler
- Astrochronology: Prof. Dr. Maria Gertrude Firneis
- Dendrochronology: Dr. Otto Cichocki
- C14-Dating: Prof. Dr. Walter Kutschera
  Co-worker: DDr. Peter Stadler
- Data Management, Electronic Communication and Quantitative Methods:
  DDr. Peter Stadler
  Co-worker: Mag. Wolfgang Müller

References

Beckerath, J. v. (1997). *Chronologie des pharaonischen Ägyptens*. Verlag Philipp von Zabern, Mainz.

Bichler, M., Egger, H., Preisinger, A., Ritter, D. and Stastny, P. (1997). NAA of the 'Minoan pumice' at Thera and comparison to alluvial pumice deposits in the Eastern Mediterranean region. *Journal of Radioanalytical and Nuclear Chemistry*, **224**, 7–14.

Bichler, M., Peltz, C., Saminger, S. and Exler, M. (2002). Aegean tephra – an analytical approach to a controversy about chronology. *Egypt and the Levant*, **XII**, 55–70.

Bietak, M. (1976). Stratigraphische Probleme bei Tellgrabungen im Vorderen Orient. Fs Pittioni. *Archaeologia Austriaca*, **14**, 471–493.

Bietak, M. (1985). Stratigraphie und Seriation. Arbeiten zur Erschließung der relativen Chronologie in Ägypten. In *Lebendige Altertumswissenschaften: Festschrift Hermann Vetters*, A. Holzhausen, Vienna, 5–9.

106   Otto Cichocki *et al.*

Bietak, M. (ed.) (2000). *The synchronisation of civilisations in the Eastern Mediterranean in the second millennium BC*. Contributions to the chronology of the Eastern Mediterranean, vol. 1. Denkschriften der Gesamtakademie, Bd. 19. Australian Academy of Sciences Press, Vienna.

Bikai, P. M. (1991). *The cedar of Lebanon: archaeological and dendrochronological perspectives*. Ph.D. thesis, University of Berkeley, Berkeley, CA, USA.

Bronk Ramsey, C. (1995). Radiocarbon calibration and analysis of stratigraphy: the OxCal program. *Radiocarbon*, **37**, 425–430.

Bronk Ramsey, C. (1998). Probability and dating. *Radiocarbon*, **40**, 461–474.

Bronk Ramsey, C. (2000). Comment on 'The use of Bayesian statistics for ^{14}C dates of chronologically ordered samples: a critical analysis'. *Radiocarbon*, **42**, 199–202.

Bronk Ramsey, C. (2001). Development of the radiocarbon calibration program OxCal. *Radiocarbon*, **43**, 355–363.

Bronk Ramsey, C., van der Plicht, J. and Weninger, B. (2001). 'Wiggle matching' radiocarbon dates. *Radiocarbon*, **43**, 381–389.

Buck, C. E., Kenworthy, J. B., Litton, C. D. and Smith, A. F. M. (1991). Combining archaeological and radiocarbon information: a Bayesian approach to calibration. *Antiquity*, **65**, 808–821.

Buck, C. E., Litton, C. D. and Scott, E. M. (1994). Making the most of radiocarbon dating: some statistical considerations. *Antiquity*, **68**, 252–263.

Buck, C. E., Litton, C. D. and Smith, A. F. M. (1992). Calibration of radiocarbon results pertaining to related archaeological events. *Journal of Archaeological Science*, **19**, 497–512.

Chapront-Touzé, M. and Chapront, J. (1991). *Lunar tables and programs from 4000 BC to AD 8000*. Willmann-Bell, Richmond.

Christen, J. A., Clymo, R. S. and Litton, C. D. (1995). A Bayesian approach to the use of ^{14}C dates in the estimation of the age of peat. *Radiocarbon*, **37**, 431–442.

Christen, J. A. and Litton, C. D. (1995). A Bayesian approach to wiggle-matching. *Journal of Archaeological Science*, **22**, 719–725.

Davies, W. V. (1995). Ancient Egyptian timber imports: an analysis of wooden coffins in the British Museum. In W. V. Davies and L. Schofield (eds.), *Egypt, the Aegean and the Levant. Interconnections in the second millennium BC*, British Museum Press.

Druitt, T. H., Edwards, L., Mellors, R. M., Pyle, D. M., Sparks, R. S. J., Lanphere, M., Davies, M. and Barriero, B. (1999). *Santorini Volcano*. Memoirs of the Geological Society of London, **19**. Geological Society, London.

Eastwood, W. J., Pearce, N. J. G., Westgate, J. A., Perkins, W. T., Lamb, H. F. and Roberts, N. (1999). Geochemistry of Santorini tephra in lake sediments from Southwest Turkey. *Global and Planetary Change*, **21**, 17–29.

Einarsson, T. (1986). Tephrochronology. In B. E. Berglund (ed.), *Handbook of Holocene palaeoecology and palaeohydrology*, John Wiley and Sons, Chichester, 392–342.

Engle, J. B. (2000). *A computer-assisted tree-ring chronology composition system*. Master's thesis, University of Arizona, Arizona, USA. URL http://www.ece.arizona.edu/~dial/xdatedoc/xdate_thesis_v3.1.pdf

Faure, P. (1971). Remarques sur la présence et l'emploi de la Pierre Ponce en Crète du Néolithique à nos jours. In X. Y. Marinatos and D. Ninkovich (eds.), *Proceedings of the 1st international scientific congress on the volcano of Thera, Athens*. Archaeological Services of Greece, 422–427.

Francaviglia, V. (1986). Provenance of pumices from the Kastelli excavations (Chania, West-Crete). In T. Hackens, Y. Liritzis, F. Dachy and G. Moucharte (eds.), *Proceedings of the first South European symposium on archaeometry, Delphi, November 1984*. PACT, Strasbourg, Journal of the European Study Group on Physical, Chemical and Mathematical Techniques Applied to Archaeology, Conseil de l'Europe, **15**, 67.

Francaviglia, V. (1990). Sea-borne pumice deposits of archaeological interest on Aegean and Eastern Mediterranean beaches. In D. A. Hardy and A. C. Renfrew (eds.), *Thera and the Aegean world III: proceedings of the third international congress, Santorini, Greece, 3–9 September 1989, Vol. 2: Earth sciences*. Thera Foundation, London, 127–134.

Francaviglia, V. and DiSabatino, B. (1990). Statistical study on Santorini pumice-falls. In D. A. Hardy and A. C. Renfrew (eds.), *Thera and the Aegean world III: proceedings of the third international congress, Santorini, Greece, 3–9 September 1989, Vol. 2: Earth sciences*. Thera Foundation, London, 29–52.

Grosser, D. et al. (1992). Holz – ein wichtiger Werkstoff im Alten Ägypten. In S. Schoske, B. Kreissl and R. Germer (eds.), *"Anch" – Blumen fÄijr das Leben. Schriften aus der ägyptischen Sammlung*, Staatliche Sammlung ägyptischer Kunst, München.

Guichard, F., Carey, S., Arthur, M. A., Sigurdsson, H. and Arnold, M. (1993). Tephra from the Minoan eruption of Santorini in sediments of the Black Sea. *Nature*, **363**, 610–612.

Kaennel, M. and Schweingruber, F. H. (1995). *Multilingual glossary of dendrochronology: terms and definitions in English, German, French, Spanish, Italian, Portuguese and Russian*. P. Haupt, Bern.

Keller, J. (1980). Prehistoric pumice tephra on Aegean islands. In D. A. Hardy, C. G. Coumas, J. A. Sakellarakis and P. M. Warren (eds.), *Thera and the Aegean world: papers presented at the second international scientific congress, Santorini, Greece, August 1978, vol. II, Thera and the Aegean world*. Thera Foundation, London, 49–56.

Kitchen, K. A. (2000). Regnal and genealogical data of Ancient Egypt: the historical chronology of Ancient Egypt, a current assessment. In M. Bietak (ed.), *The synchronisation of civilisations in the eastern Mediterranean in the second millennium BC*. Australian Academy of Sciences Press, Vienna, 39–52.

Kuniholm, P. I. (1996). The prehistoric Aegean: dendrochronological progress as of 1995. *Acta Archaeologica*, **67**, 327–335.

Lev-Yadun, S. (1992). The origin of the cedar beams from Al-Aqsa Mosque: botanical, historical and archaeological evidence. *Levant*, **24**, 201–208.

Library, M. (1983). *The Macquarie history of ideas*. Macquarie University, Sydney.

Liphschitz, N. (1986). Overview of the dendrochronological and dendroarcheological research in Israel. *Dendrochronologia*, **4**, 37–58.

Liphschitz, N. and Biger, G. (1992). Israel: historical timber trade in the Levant: the use of *Cedrus libani* in construction of buildings in Israel from ancient times to the early twentieth century. In T. S. Bartholin et al. (eds.), *Tree rings and environment*, Lund University, Department of Quaternary Geology, Lund, 202–206.

Loret, V. (1916). Quelques notes sur l'arbre « ach ». *Annales du service des antiquités de l'Égypte*, **16**, 33–51.

Lucas, U. and Harris, J. R. (1962). *Ancient Egyptian materials and industries*. Edward Arnold, London.

Maguire, L. (1992). A cautious approach to the Middle Bronze Age chronology of Cyprus. *Ägypten und die Levante*, **3**, 115–120.

Maguire, L. (1995). Tell el Dab'a: the Cypriote Connection. In W. V. Davies and L. Schonfield (eds.), *Egypt, the Aegean and the Levant*, British Museum Press, London, 54–65.

Manning, S. W. (1999). *A test of time: the volcano of Thera and the chronology and history of the Aegean and East Mediterranean in the mid second millennium*. Oxbow Books, Oxford.

Marinatos, S. (1932). Die Ausgrabungen von Amnisos auf Kreta. *Praktika tes en Athenais Archaiologikes Hetaireias*, **1932**, 76–95.

Marketou, T. (1990). Santorini tephra from Rhodes and Kos: some chronological remarks based on the stratigraphy. In D. A. Hardy and A. C. Renfrew (eds.), *Thera and the Aegean world III: proceedings of the third international congress, Santorini, Greece, 3–9 September 1989, Vol. 3: Chronology*. Thera Foundation, London, 100–113.

Meiggs, R. (1982). *Trees and timber in the ancient Mediterranean world*. Clarendon, Oxford.

Mikesell, M. (1969). The deforestation of Mount Lebanon. *Geographical Review*, **59**, 1–28.

Newton, M. W. (1996). *Dendrochronology at Catal Hoyuk: A 576-year tree ring chronology for the early Neolithic of Anatolia*. Master's thesis, Cornell University, Ithaca, NY, USA.

Nibbi, A. (1981). *Ancient Egypt and some eastern neighbours*. Noyes Press, Park Ridge, New Jersey.

Nibbi, A. (1987). Some remarks on the LEXIKON ENTRY: Zeder, Cedar. *Discussions in Egyptology*, **7**, 13–27.

Nibbi, A. (1994). Some remarks on the Cedar of Lebanon. *Discussions in Egyptology*, **28**, 35–52.

Nibbi, A. (1996). Cedar again. *Discussions in Egyptology*, **34**, 37–59.

North, J. (1994). *The Fontana history of astronomy and cosmology*. Fontana Press, London.

O'Neil, W. M. (1986). *Early astronomy*. University Press Sydney, Sydney.

Paul, R. (1993). *A handbook to the universe*. Chicago Review Press, Chicago.

Pearce, N. J. G., Westgate, J. A., Perkins, W. T., Eastwood, W. J. and Shane, P. (1999). The application of laser ablation ICP-MS to the analysis of volcanic glass shards from tephra deposits: bulk glass and single shard analysis. *Global and Planetary Change*, **21**, 151–171.

Peltz, C., Schmid, P. and Bichler, M. (1999). INAA of Aegean pumices for the classification of archaeological findings. *Journal of Radioanalytical and Nuclear Chemistry*, **242**, 361–377.

Pichler, H. and Schiering, W. (1980). Der spätbronzezeitliche Ausbruch des Thera-Vulkans und seine Auswirkungen auf Kreta. *Archäologische Anzeiger*, **1980**, 1–37.

Pyle, D. M. (1990). New estimates for the volume of the Minoan eruption. In D. A. Hardy and A. C. Renfrew (eds.), *Thera and the Aegean world III: proceedings of the third international congress, Santorini, Greece, 3–9 September 1989, Vol. 2: Earth sciences*. Thera Foundation, London, 113–120.

Richardson, D. and Ninkovich, D. (1976). Use of K$_2$O, Rb, Zr and Y versus SiO$_2$ in volcanic ash layers of the eastern Mediterranean to trace their source. *Geological Society of America Bulletin*, **87**, 110–116.

Richmond, B. (1956). *Time measurement and calendar construction*. E. J. Brill, Leiden.

Ryholt, K. S. B. (1997). *The political situation in Egypt during the Second Intermediate Period c. 1800–1550 BC*. CNI Publications, **20**. Carsten Niebuhr Institute of Near Eastern Studies, University of Copenhagen, Museum Tusculanum Press, Copenhagen.

Saminger, S., Peltz, C. and Bichler, M. (2000). South Aegean volcanic glass – separation and analysis by INAA and EPMA. *Journal of Radioanalytical and Nuclear Chemistry*, **245**, 375–383.

Schaefer, B. (1997). Heliacal risings: definitions, calculations and some specific cases. *Archaeoastronomy and Ethnoastronomy News*, **25**, 1 and 3.

Schmid, P., Peltz, C., Hammer, V. M. F., Halwax, E., Ntaflos, T., Nagl, P. and Bichler, M. (2000). Separation and analysis of Theran volcanic glass by INAA, XRF and EPMA. *Mikrochimica Acta*, **133**, 143–149.

Sigurdsson, H., Carey, S. and Devine, D. (1990). Assessment of mass, dynamics and environmental effects of the Minoan eruption of Santorini Volcano. In D. A. Hardy and A. C. Renfrew (eds.), *Thera and the Aegean world III: proceedings of the third international congress, Santorini, Greece, 3–9 September 1989, Vol. 2: Earth sciences*. Thera Foundation, London, 100–112.

110   Otto Cichocki *et al.*

Soles, J. S. and Davaras, C. (1990). Theran ash in Minoan Crete: new excavations in Mochlos. In D. A. Hardy and A. C. Renfrew (eds.), *Thera and the Aegean world III: proceedings of the third international congress, Santorini, Greece, 3-9 September 1989, Vol. 3: Chronology*. Thera Foundation, London, 89-95.

Sparks, R. and Wilson, C. (1990). The Minoan deposits: a review of their characteristics and interpretation. In D. A. Hardy and A. C. Renfrew (eds.), *Thera and the Aegean world III: proceedings of the third international congress, Santorini, Greece, 3-9 September 1989, Vol. 3: Chronology*. Thera Foundation, London, 89-99.

Stanley, D. J. and Sheng, H. (1986). Volcanic shards from Santorini (Upper Minoan ash) in the Nile Delta, Egypt. *Nature*, **320**, 733-735.

Sullivan, D. G. (1988). The discovery of Santorini Minoan tephra in western Turkey. *Nature*, **333**, 552-554.

Vinci, A. (1985). Distribution and chemical composition of tephra layers from eastern Mediterranean abyssal sediments. *Marine Geology*, **64**, 143-155.

Vitaliano, C. J., Fout, J. S. and Vitaliano, D. B. (1978). Petrochemical study of the tephra sequence exposed in the Phira Quarry, Thera. In D. A. Hardy, C. G. Coumas, J. A. Sakellarakis and P. M. Warren (eds.), *Thera and the Aegean world: papers presented at the second International Congress, Santorini, Greece, August, 1978, Vol. I, Thera and the Aegean world*. Thera Foundation, London, 203-215.

Vitaliano, C. J., Taylor, S. R., Norman, M. D., McCulloch, M. T. and Nicholls, I. A. (1990). Ash layers of the Thera volcanic series: stratigraphy, petrology and geochemistry. In D. A. Hardy and A. C. Renfrew (eds.), *Thera and the Aegean world III: proceedings of the third international congress, Santorini, Greece, 3-9 September 1989, Vol. 3: Chronology*. Thera Foundation, London, 53-78.

Wagenführ, R. and Schreiber, C. (1989). *Holzatlas*. VEB Fachbuchverlag, Lepizig.

Warren, P. M. and Puchelt, H. (1990). Stratified pumice from Bronze Age Knossos. In D. A. Hardy and A. C. Renfrew (eds.), *Thera and the Aegean world III: proceedings of the third international congress, Santorini, Greece, 3-9 September 1989, Vol. 3: Chronology*. Thera Foundation, London, 71-81.

Watkins, N. D., Sparks, R. S. J., Sigurdsson, H., Huang, T. C., Federman, A., Carey, S. and Ninkovich, D. (1978). Volume and extent of the Minoan tephra from Santorini Volcano: new evidence from deep-sea sediment cores. *Nature*, **271**, 122-126.

Wild, E., Golser, R., Hille, P., Kutschera, W., Priller, A., Puchegger, S., Rom, W., Steier, P. and Vycudilik, V. (1998). First 14C results from archaeological and forensic studies at the Vienna Environmental Research Accelerator. *Radiocarbon*, **40**, 273-281.

Zeist, V. V., Woldring, H. and Stapert, D. (1975). Late Quaternary vegetation and climate of southwestern Turkey. *Paleohistoria*, **17**, 53-143.

5

Applications of Formal Model Choice to Archaeological Chronology Building

Sujit K. Sahu

Summary. This chapter provides an overview of a topic that is likely to become increasingly important as greater numbers of researchers adopt formal statistical models for constructing chronologies. Other chapters in this volume (1, 2, 3, 10 and 11) use single statistical models, but in the future, as researchers attempt to draw together coherently information from different sources, they will almost certainly develop several alternative models for a single problem. Different statistical models may, however, produce very different interpretations of the same data and thus give rise to conflicting reconstructions of the past. In such situations, we need a robust way to investigate which models are best supported by the data. This chapter outlines recent developments in the application of formal Bayesian model choice techniques to archaeological chronology building and illustrates these tools using two examples, one from absolute and the other from relative chronology building problems. A particular advantage of Bayesian model choice techniques lies in their ability to compare widely different models based on differing assumptions and prior information.

5.1 Introduction

Statistical methods are now an essential part of the archaeological inference making process as illustrated in the books by Shennan (1998), Baxter (1994) and Buck et al. (1996). The statistical techniques I discuss here are based on the Bayesian paradigm which provides a natural and convenient way to incorporate prior information in practical problems. Although some authors criticize the Bayesian view (see for example Reece 1994), it is regarded as *the* most general and coherent statistical inference procedure capable of solving practical problems. Bayesian methods have many advantages, see for example the book by Buck et al. (1996). One particular advantage lies in their ability to compare widely different models based on different assumptions and prior information.

Prior information, although quite valuable, cannot build chronologies for *sure*. Often such information comes from expert archaeologists working on particular problems of interpretation of archaeological data. However, experts often disagree and they may provide different archaeological dates or explanations of the data. Moreover, the adopted statistical models are also liable to be uncertain. An evaluation of model uncertainties is required before making the final inference. Thus in practical problems both the prior information and the assumed statistical models need to be thoroughly examined as part of the model evaluation process. The Bayesian methods I am going to describe allow us to address and measure the uncertainties arising due to the possible mis-specification of the statistical model for the data and the assumed prior distribution for the parameters.

Archaeologists often seek two types of chronological evidence: absolute and relative. Absolute techniques provide estimates of the true calendar date of archaeological events. Relative techniques, on the other hand, simply allow estimates of the chronological order in which events took place. Of course, if absolute dates were available for all events of interest, relative dating would not be needed. Typically, however, this is not the case and ways are sought to combine both relative and absolute chronological information in order to enhance temporal understanding.

In this chapter I shall illustrate the model choice methods with a simple theoretical example and two practical examples. The first practical example is taken from Nicholls and Jones (2001). It compares two alternative prior distributions for the boundary parameters that divide excavated layers in a chronological model they adopted for the purposes of absolute dating. Using one set of prior assumptions for this model, they obtain a much tighter posterior distribution for the span of the absolute dates than that obtained using another, widely adopted, prior. Bayesian model choice methods help decide between the two prior models which give rise to two completely different posterior distributions. The interpretations obtained from these distributions are also quite different and hence one must choose a model from the two alternatives considered.

The second practical example concerns relative dating of artefacts or excavated materials. On some archaeological excavations there are no reliable relationships between vertical location in the ground and relative date of deposition of the artefacts found. In some others, as in the following example, it is not possible to link excavated material on one site with those at another. Statistical methodologies based upon the artefacts excavated are sometimes employed in an attempt to derive relative chronological information. Such methodologies, which identify temporal sequence on the basis of the number of different types of artefacts, are commonly referred to as *seriation* techniques.

Buck and Sahu (2000) consider seriation of a data set relating to the numbers of seven types of mesolithic flint tools (known as microliths) from six different sites, numbered 1 to 6, in southern England (Jacobi et al. 1980). The objective is to identify the relative chronological order of the sites by studying the changes in the numbers of the seven types of microliths found at them. Two widely used competing methodologies suggest completely different orders for the sites: 2, 5, 3, 6, 1, 4 and 3, 6, 5, 2, 1, 4. Clearly these are likely to give rise to quite different archaeological conclusions. The problem I focus on here is to choose between the two using model choice methods.

The remainder of this chapter is organized as follows. Section 5.2 provides the *model choice* framework within which statistical solutions are proposed. The model choice criteria are illustrated using a simple theoretical example in Section 5.3. Further, a radiocarbon dating example is provided in Section 5.4 and an example on relative chronology building is discussed in Section 5.5. Finally, a few summary remarks are made in Section 5.6.

5.2 Bayesian Methods

5.2.1 MCMC Model Fitting

Currently Markov chain Monte Carlo (MCMC) simulation techniques are used in a wide variety of statistical problems with relative ease and great success. These methods allow critical re-examination of existing model-based approaches and are flexible enough to posit and develop more realistic models.

Let $\mathbf{y}$ denote the observed data to be modelled and let ζ denote the unknown parameters in the model. Let $\pi(\zeta|\mathbf{y})$ denote the posterior distribution of the parameters ζ under the assumed Bayesian model. In order to implement the MCMC method known as the Gibbs sampler (Gelfand and Smith 1990) one writes down the complete conditional posterior distribution of all the parameters. These distributions have densities which are all proportional to the joint posterior density $\pi(\zeta|\mathbf{y})$. The Gibbs sampler then simulates from each conditional distribution in turn for a large number of times, B say, starting from an arbitrary point. For large values of B, the effect of the starting point is forgotten and one obtains random samples from the joint posterior distribution. Features of the posterior distribution are then estimated accurately using appropriate averages of samples so obtained. For a general introduction to MCMC methods see the book by Gilks et al. (1996).

Due to the complexity of archaeological problems, however, many authors have shown that some ingenuity is needed in devising sampling schemes. Buck and Sahu (2000), for example, document several different attempts at implementation before a successful sampling scheme was devised. Once efficient algorithms for fitting statistical models for large and complex archaeological data sets have been implemented then the validity of the fitted models should be checked. I propose to use predictive Bayesian model choice techniques both to facilitate model comparison and assess goodness-of-fit. Bayesian model checking serves the latter purpose and is important because in model selection we run the risk of selecting from a set of badly fitting alternatives.

5.2.2 Predictive Distributions

Bayesian model choice methods can be based on predictive distributions. In simple terms, these are distributions of future replicate data sets obtained by eliminating the parameter uncertainties. Different types of predictive distributions arise by considering different methods of eliminating the uncertain parameters. A few predictive distributions are listed below. These distributions will be used subsequently to define model choice criteria.

Let y_{obs} denote the observed data with individual data points $y_{r,obs}, r = 1, \ldots, n$, and y_{rep} with components $y_{r,rep}$ (abbreviation for replicate) denote a future set of observables under the assumed model. Further, let $y_{(r),obs}$ denote the set of observations y_{obs} with rth component deleted.

The prior predictive distribution of a set of observations at the actual observed point y_{obs} has the density given by

$$\pi(y_{obs}) = \int \pi(y_{obs}|\zeta)\, \pi(\zeta)\, d\zeta. \tag{5.1}$$

In the Bayesian inference setup the actual observations y_{obs} are fixed, and the above is interpreted as the density of a set of observables evaluated at the observed point y_{obs}. This is also known as the *marginal likelihood* of the data. The prior predictive density is only meaningful if the prior distribution $\pi(\zeta)$ is a proper distribution (i.e. $\int \pi(\zeta)d\zeta = 1$), due to its involvement in the definition in Equation 5.1.

The *cross-validation predictive density* is defined by

$$\pi(y_r|y_{(r),obs}) = \int \pi(y_r|\zeta, y_{(r),obs})\, \pi(\zeta|y_{(r),obs})\, d\zeta. \tag{5.2}$$

In the case of conditionally independent observations given ζ,

$$\pi(y_r|\zeta, y_{(r),obs}) = \pi(y_r|\zeta).$$

The predictive density in Equation 5.2 then simplifies to

$$\pi(y_r|y_{(r),obs}) = \int \pi(y_r|\zeta)\, \pi(\zeta|y_{(r),obs})\, d\zeta. \tag{5.3}$$

This density is also known as the *conditional predictive ordinate* (CPO). These densities are meaningful even when improper prior distributions for ζ are considered as long as the posterior distribution $\pi(\zeta|y_{(r),obs})$ is proper for each r.

The *posterior predictive density* of y_{rep}, given by

$$\pi(y_{rep}|y_{obs}) = \int \pi(y_{rep}|\zeta)\, \pi(\zeta|y_{obs})\, d\zeta, \tag{5.4}$$

is the predictive density of a new independent set of observables y_{rep} under the model, given the actual data y_{obs}. The posterior predictive density is easier to work with than the previous two densities (i.e. Equations 5.1 and 5.2), because features of $\mathbf{y}_{\text{rep}}$ having the density in Equation 5.4 can be estimated easily when MCMC samples from the posterior $\pi(\zeta|\mathbf{y}_{\text{obs}})$ are available. A new set of observations drawn from $\pi(\mathbf{y}_{\text{rep}}|\zeta)$, the likelihood model conditional on ζ, is a sample from the predictive density in Equation 5.4.

5.2.3 The Bayes Factor

A pure Bayesian approach to model selection is to report posterior probabilities of each model by comparing Bayes factors, see for example DiCiccio et al. (1997) and Kass and Raftery (1995). The *Bayes factor* (BF) for comparing two given models M_1 and M_2 is

$$\text{BF} = \frac{\pi(\mathbf{y}_{\text{obs}}|M_1)}{\pi(\mathbf{y}_{\text{obs}}|M_2)},$$

where $\pi(\mathbf{y}_{\text{obs}}|M_i)$ is the density in Equation 5.1 when M_i is the assumed model, $i = 1, 2$.

The BF gives a summary of the evidence for M_1 against M_2 provided by the data. Calibration tables for the BF are available for deciding how strong the evidence is, see for example Kass and Raftery (1995). Recall that $\pi(\mathbf{y}_{\text{obs}}|M_i)$ is the marginal likelihood of the data under model M_i. Hence the BF chooses a model for which the marginal likelihood of the data is maximum.

The cross-validation predictive densities are used to form a variant of the Bayes factor called the *pseudo-Bayes factor* (PsBF) (Geisser and Eddy 1979). For comparing two models M_1 and M_2 the PsBF is defined as,

$$\text{PsBF} = \prod_{r=1}^{n} \frac{\pi(y_r|\mathbf{y}_{(r),\text{obs}}, M_1)}{\pi(y_r|\mathbf{y}_{(r),\text{obs}}, M_2)}.$$

This is a surrogate for the Bayes factor and its interpretations are similar, see for example Gelfand (1996). The CPOs are also useful for checking model adequacy. Instead of using a single summary measure alone, e.g. the PsBF, the individual CPOs can also be compared under any two models. This is to guard against any single highly influential observation concealing a general trend. One observation, $y_{r,\text{obs}}$, prefers model M_1 to M_2 if the rth CPO is higher under M_1. The CPOs are not illustrated in this chapter since the primary issue here is model choice and not model checking.

5.2.4 A Decision-Theoretic Approach

Gelfand and Ghosh (1998) and Laud and Ibrahim (1995) propose model selection criteria based on the posterior predictive densities. The current model is a 'good' fit to the observed data, $\mathbf{y}_{\text{obs}}$, if $\mathbf{y}_{\text{rep}}$ is able to replicate the data well. Hence, many model choice criteria can be developed by considering different loss functions for measuring the divergence between $\mathbf{y}_{\text{obs}}$ and $\mathbf{y}_{\text{rep}}$ (see for example Rubin 1984). If the data are assumed to be symmetrically distributed with a common variance then it is mathematically convenient to adopt a squared error loss function

$$L(\mathbf{y}_{\text{rep}}, \mathbf{y}_{\text{obs}}) = \sum_r (y_{r,\text{rep}} - y_{r,\text{obs}})^2. \tag{5.5}$$

In the unequal variance case one may weight the individual terms in the loss function by the inverse variance of $y_{r,\text{obs}}$ if it is known. Other loss functions are also possible, for example Buck and Sahu (2000) use the following deviance loss function

$$L(\mathbf{y}_{\text{rep}}, \mathbf{y}_{\text{obs}}) = 2 \left(\sum_r y_{r,\text{obs}} \log \frac{y_{r,\text{obs}}}{y_{r,\text{rep}}} \right) \tag{5.6}$$

where the data are assumed to follow the multinomial distribution. The best model among a given set of models is the model for which the expected value of the adopted loss function is the minimum, where the expectation is to be taken with respect to the posterior predictive distribution in Equation 5.4.

5.2.5 The Deviance Information Criterion (DIC)

There are many other Bayesian methods available for model comparison. These methods use the posterior distribution of the likelihood to arrive at suitable model choice criteria. For example, Aitkin (1997) interprets the p-values by using the posterior distribution of the likelihood function.

Recently, Spiegelhalter et al. (2002) proposed a model selection criterion for arbitrarily complex models called the deviance information criterion (DIC). They first define the deviance function as

$$D(\zeta) = -2 \log\{\pi(\mathbf{y}|\zeta)\} + 2 \log\{\pi(\mathbf{y})\}$$

where $\pi(\mathbf{y}|\zeta)$ is the likelihood function, and $\pi(\mathbf{y}) = \pi(\mathbf{y}|\mu(\zeta) = \mathbf{y})$. Here $\mu(\zeta)$ is defined as the mean of the data, that is $\mu(\zeta) = E(\mathbf{Y}|\zeta)$.

The DIC is defined to be the sum of two components: the deviance evaluated at the posterior mean and a penalty factor which penalizes for model complexity. The penalty factor, denoted by p_D, is defined as

$$p_D = E\{D(\zeta)|\mathbf{y}\} - D\{E(\zeta|\mathbf{y})\}.$$

Thus p_D is the expected deviance minus the deviance evaluated at the posterior expectations. The p_D is called the *effective number of parameters* in a complex model. Subsequently, they define the model choice criterion

$$\text{DIC} = D\{E(\zeta|\mathbf{y})\} + 2\,p_D.$$

The model with the smallest DIC is chosen to be the best model for data.

5.3 A Simple Example

I first consider a simple example which reveals the Bayesian model choice criteria in closed form analytic expressions. Suppose that $y_1, \ldots, y_n$ are observations from the $N(\theta, 1)$ population and the prior for θ is $N(0, \tau^2)$ where τ^2 is known and finite. Thus, in this example, $\theta = \zeta$. Consider the two models:

$$M_1 : \theta = 0, \quad \text{vs} \quad M_2 : \theta \neq 0.$$

This is perhaps overly simplified, but the setup will aid understanding of the Bayesian model choice criteria.

The posterior distribution of θ is given by

$$\pi(\theta|\mathbf{y}) = N\left(\frac{n\bar{y}}{n + 1/\tau^2}, \frac{1}{n + 1/\tau^2}\right).$$

Thus the observations $y_1, \ldots, y_n$ enter into the posterior distribution through $\bar{y}$ and the implied model for the data is

$$\bar{Y} = \frac{1}{n}\sum Y_i \sim N\left(\theta, \frac{1}{n}\right).$$

This also means that $\bar{Y}$ is the sufficient statistic for θ.

Suppose that Z is a future observation for which I wish to calculate the predictive distribution. To have simpler notation I let $Z = \mathbf{y}_{\text{rep}}$ and $\mathbf{y}_{\text{obs}} = \bar{y}$. If θ is known I have, $\pi(Z|\theta) = N\left(\theta, \frac{1}{n}\right)$. The predictive distribution (prior or posterior) of Z has two different forms under the two models, M_1 and M_2. Under model M_1 there are no unknown parameters and both the prior and posterior predictive distributions are given by $N\left(0, \frac{1}{n}\right)$.

The prior predictive distribution (Equation 5.1) of Z under model M_2 is given by

$$\pi(z) = N\left(0, \frac{1}{n} + \tau^2\right).$$

As expected, if $\tau^2 = 0$ this distribution reduces to the prior predictive under M_1. The posterior predictive distribution is calculated as

$$\pi(z|\bar{y}) = N\left(\frac{n\tau^2}{n\tau^2 + 1}\bar{y}, \frac{1}{n} + \frac{\tau^2}{n\tau^2 + 1}\right).$$

As expected this posterior predictive distribution has less variability than the prior predictive distribution for non-zero values of τ^2. This fact will be further discussed in Section 5.4. Moreover, the centre of the distribution is located near the centre of the data $\bar{y}$, unlike the prior predictive distribution which is centred at zero, the prior mean. The cross-validation predictive distributions are not considered because effectively there is only one data point $\bar{y}$.

Assume the loss function to be

$$L(z, \bar{y}) = (z - \bar{y})^2.$$ (5.7)

Now, the following decision rules are derived based on the three predictive model selection criteria. Select model M_1 if

$$n\bar{y}^2 < (1 + n\tau^2) \frac{\log(1 + n\tau^2)}{n\tau^2}, \text{ using the Bayes factor,}$$
$$< (1 + n\tau^2) \frac{1}{2 + n\tau^2}, \text{ using the squared error loss function, Equation 5.7}$$
$$< (1 + n\tau^2) \frac{2}{2 + n\tau^2}, \text{ using the DIC.}$$

It can be verified that

$$\frac{\log(1 + n\tau^2)}{n\tau^2} \geq \frac{2}{2 + n\tau^2} > \frac{1}{2 + n\tau^2}.$$

The above results are interpreted as follows. If the loss function based approach selects model M_1 then the Bayes factor will select the same as well. The loss function based approach is likely to reject the simpler model M_1 more often than the Bayes factor based approach. The last two predictive criteria criticize the simpler model too much. They require the models to both fit and predict the data well. Thus the Bayes factor is seen to be less stringent regarding the choice of the simpler model than the remaining two model choice criteria. I shall extrapolate this theoretical result for the practical example in Section 5.5.

5.4 Example: Absolute Chronology Building

5.4.1 Models

As indicated by several other chapters in this volume (in particular Chapters 1, 2 and 9) absolute chronologies are often built using radiocarbon data. Typically, radiocarbon dating laboratories provide a CRA (conventional radiocarbon age) and an estimate of the associated error for a given sample from a dead organism. Statistical methods together with internationally agreed high-precision calibration data are then used to convert the CRA to usable calendar dates (for details see Chapter 1) A full set of calibration data is available from http://depts.washington.edu/qil/ (see also Stuiver et al. 1998).

By way of an example, consider a set of seven CRA determinations and associated errors which is a subset of a large set of data gathered at the mouth of the Shag River, in southern New Zealand. The data set, given in Table 5.1, consists of all charcoal dates from a single series of six layers. Interest here focuses on the actual dates of deposition of the samples and the length of time for which the site was occupied.

I now discuss statistical formulation of the above problem as described by Nicholls and Jones (2001). Here the problem is simultaneously to calibrate

Table 5.1. Seven conventional radiocarbon determinations from samples gathered at the mouth of the Shag River in southern New Zealand.

Layer	Sample	CRA	Lab sd	Sample ID
1	1	580	47	NZ 7758
2	1	600	50	NZ 7761
3	1	537	44	NZ 7757
4	1	670	47	NZ 7756
5	1	646	47	NZ 7755
5	2	630	35	WK 2589
6	1	660	46	NZ 7771

several radiocarbon determinations found in a vertical series of a number of abutting layers of earth, I say. Suppose n_i CRA determinations are available from layer i, making $n = \sum_{i=1}^{I} n_i$ data points in all. Let Y_{ij} denote the value of the jth CRA measured in the ith layer and let θ_{ij} denote the corresponding true calendar date. Associated with θ_{ij} is a unique radiocarbon age, $\mu(\theta_{ij})$, which relates to the amount of ^{14}C present in the sample when it is measured. It is often assumed that

$$Y_{ij} = \mu(\theta_{ij}) + \epsilon_{ij}^{(Y)} + \epsilon_{ij}^{(\mu)}, \tag{5.8}$$

where $\epsilon_{ij}^{(Y)}$ and $\epsilon_{ij}^{(\mu)}$ are independent normal random variables with zero means. Let the variance of $\epsilon_{ij}^{(Y)}$ be

$$\sigma_{ij}^{(Y)^2}$$

and the variance of $\epsilon_{ij}^{(\mu)}$ given θ_{ij} be

$$\sigma^{(\mu)^2}(\theta_{ij}).$$

Thus, given θ_{ij}, the variance of Y_{ij} is

$$\sigma_{ij}^{(Y)^2} + \sigma^{(\mu)^2}(\theta_{ij}).$$

The quantities $\mu(\theta_{ij})$ and $\sigma^{(\mu)}(\theta_{ij})$ are obtained using piece-wise linear functions of calibration data. As discussed by Buck (Chapter 1), there are other methods of determining the calibration functions $\mu(\theta)$ and $\sigma^{(\mu)}(\theta)$; for example, using Gaussian process prior models (Gómez Portugal Aguilar et al. 2002).

Following Nicholls and Jones (2001), in my setup vertical mixing of earth is assumed to occur within layers, but not between layers. Let ψ_i denote the calendar date associated with the boundary between layers i and $i+1$, $i = 0, 1, \ldots, I$. Moreover, assume that layer $i = 1$ is the topmost and most recent layer, while layer $i = I$ is the deepest layer containing the oldest material.

Let P and A $(P \leq A)$ denote the lower and upper bounds on the unknown parameters ψ. In any layer i, the n_i calendar dates, $\theta_i = (\theta_{i1}, \dots, \theta_{in_i})$ are all assumed to be in the interval (ψ_i, ψ_{i-1}). No other constraints are put on the calendar dates within a layer. Thus the model parameters satisfy the stratigraphic constraints

$$P \leq \psi_I \leq \theta_I \leq \psi_{I-1} \leq \theta_{I-1} \leq \dots \leq \psi_0 \leq A,$$

where the inequalities hold element-wise. Based on the constraints, and without any other more specific prior information, it is reasonable to assume independent uniform prior distributions for each component of θ_i in the interval (ψ_i, ψ_{i-1}). Thus the prior distribution for θ conditional on ψ is

$$\pi(\theta|\psi) = \prod_{i=1}^{I} \prod_{j=1}^{n_i} (\psi_{i-1} - \psi_i)^{-1} I(\psi_i \leq \theta_{ij} \leq \psi_{i-1}),$$

where $I(\cdot)$ is the indicator function. To complete the prior specification it remains to consider suitable prior distributions for the boundary parameters ψ.

As outlined by Buck (Chapter 1), a widely adopted prior distribution seeks to represent prior ignorance on the relative positions of individual ψ_is. It assumes that the unordered ψs follow the uniform distribution in the interval (P, A). The ordered samples, in increasing order, are then taken as the ψ parameters. Suppose that $U_0, U_1, \dots, U_I$ is a random sample from the uniform distribution in the interval (P, A), then I set $\psi_i = U_{(i)}$ where $U_{(I)} \leq U_{(I-1)} \leq U_{(0)}$. The associated prior distribution for ψ has the prior density

$$\pi^{(1)}(\psi) = \frac{(I+1)!}{R^{I+1}}, \quad P \leq \psi_I \leq \dots \leq \psi_0 \leq A.$$

Nicholls and Jones (2001) suggest an alternative prior distribution which they call the reference prior distribution. They assume that the span $\delta = \psi_0 - \psi_I$ follows the uniform distribution in the interval $(0, R)$ where $R = A - P$. Given δ, $\psi_I \sim U(P, A - \delta)$. This defines a joint prior distribution for the two endpoints, ψ_I and ψ_0. Given the two endpoints, the remaining $(I - 1)$ unordered boundary parameters are assumed to follow the uniform distribution in the interval (ψ_I, ψ_0) independently. Suppose that $U_1, \dots, U_{I-1}$ is a random sample from the uniform distribution in the interval (ψ_I, ψ_0), then I set, $\psi_i = U_{(i)}, i = 1, \dots, I - 1$ where $U_{(I-1)} \leq \dots \leq U_{(1)}$. The prior density of ψ is

$$\pi^{(2)}(\psi) = \frac{(I-1)!}{(\psi_0 - \psi_I)^{I-1}} \frac{1}{R(R - \psi_0 + \psi_M)}, \quad P \leq \psi_I \leq \dots \leq \psi_0 \leq A.$$

There are fundamental differences between the two prior distributions $\pi^{(1)}$ and $\pi^{(2)}$, see Figure 5.1 where these densities are plotted. Under $\pi^{(1)}$ the span,

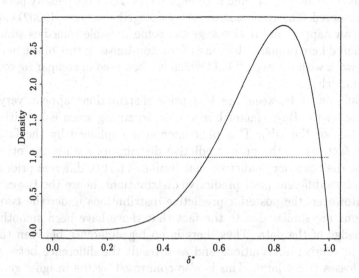

Fig. 5.1. Prior densities for δ^*. Solid line is the density under $\pi^{(2)}$ and dotted line is the density under $\pi^{(1)}$.

δ, is the distribution of $U_{(0)} - U_{(I)}$ where $U_0, U_1, \ldots, U_I$ are random samples from the uniform distribution in the interval (P, A). As a result the density of δ is the density of the sample range where the sample is obtained from the uniform distribution. The density of δ is given by

$$\pi^{(1)}(\delta) = \frac{I(I+1)}{R^{I+1}} \delta^{I-1}(R - \delta), \quad 0 < \delta < R.$$

That is, $\delta^* = \delta/R$ follows the standard beta distribution with parameters I and 2. However, under $\pi^{(2)}$, δ follows the uniform distribution in $(0, R)$, consequently δ^* follows the uniform distribution in $(0, 1)$. Nicholls and Jones (2001) point out that $\pi^{(1)}$ is more informative about δ than $\pi^{(2)}$.

Nicholls and Jones formulate this problem as a Bayesian model choice problem and use the Bayes factor to decide between the prior distributions. Here I shall compare the two prior models using the predictive Bayesian methods discussed earlier.

5.4.2 Results

The Bayes factor for comparing model 2 (with prior $\pi^{(2)}$) against model 1 (with prior $\pi^{(1)}$) has been reported to be 26 by Nicholls and Jones (2001). However, it is interesting to see what would have happened had they used the other model choice criteria described earlier.

In Table 5.2 I report all three model choice criteria for the two models. Both the expected loss criterion and the DIC criterion choose model 2 as well. For the non-linear models considered here *negative* values of the penalty parameter p_D were observed. The discussion paper by Spiegelhalter et al. (2002) explains why this can happen and it also suggests some possible remedies which are not considered here primarily because of the non-linear nature of the models. I choose to work with the overall DIC which is often used in comparing complex Bayesian models.

The differences between the two prior distributions appear very pronounced under the Bayes factor but are not so strong when using either the expected loss or the DIC. This phenomenon is explained by the fact that the Bayes factor uses the prior predictive distributions while the other two criteria use the posterior predictive distribution. The two different prior distributions induce different prior predictive distributions, hence the Bayes factor is large. However, the posterior predictive distributions under the two prior distributions are similar due to the fact that those have been smoothed by the knowledge of the data. Thus there is no big difference between the two posterior predictive distributions and as a result the difference between the expected losses is not large. This is also confirmed by the insights gained in the simple theoretical example in Section 5.3. There it is seen that the posterior predictive distribution is smoother (i.e. has less variability) than the prior predictive distribution.

Table 5.2. Model choice for the Shag River data.

	Expected loss	DIC	BF
Model 1	13.97	4.51	1
Model 2	12.51	1.17	26

5.5 Example: Relative Chronology Building

Return to the relative chronology building example mentioned in the introduction. Buck and Sahu (2000) have developed the loss function based model choice method for this example. Here I experiment with the other criteria as well.

5.5.1 The Robinson–Kendall Model

Consider the following extension of the model originally proposed by Kendall (1971). Let y_{ij} denote the observed number of artefacts (for example, pottery or tool types) of type j ($j = 1, \ldots, J$) found at archaeological site i ($i = 1, \ldots, I$). Let $N = \sum_{ij} y_{ij}$ denote the total number of artefacts. Also, let θ_{ij} denote the underlying proportion of artefact j available for deposition at

i and let Θ denote the matrix with elements θ_{ij}. Let $\boldsymbol{\theta}$ denote the vector representation of Θ. Since θ_{ij}s are proportions it is assumed that $\theta_{ij} \geq 0$ and $\sum_i \sum_j \theta_{ij} = 1$. The problem then, is to estimate the true temporal order of the I rows which is a permutation of the indices $1, \ldots, I$. The true permutation is represented using $p(1), p(2), \ldots, p(I)$.

Assume that $\mathbf{y}$ has a multinomial distribution with parameters N and $\boldsymbol{\theta}$. That is, the probability of obtaining the observed configuration y_{ij} is given by

$$N! \prod_{ij} \theta_{ij}^{y_{ij}} / y_{ij}!. \tag{5.9}$$

A suitable way to represent prior information about θ_{ij} is to use the Dirichlet distribution, thus

$$\pi(\boldsymbol{\theta}) \propto \prod_{ij} \theta_{ij}^{\alpha_{ij}-1},$$

where $\alpha_{ij} > 0$ for all i and j. Note that I can use this to incorporate both informative and non-informative prior information quite successfully. For example, setting $\alpha_{11} = 5$ can be thought of as having four artefacts of type 1 in row 1. To specify non-informative prior information simply set $\alpha_{ij} = 0.5$ for all i and j. The posterior distribution of $\boldsymbol{\theta}$ is then given by

$$\pi(\boldsymbol{\theta}) \propto \prod_{ij} \theta_{ij}^{y_{ij}+\alpha_{ij}-1}. \tag{5.10}$$

This posterior distribution will be used to make inference about the orders.

Suppose that the true chronological order is the given natural order, i.e. $p(1) = 1$, $p(2) = 2$, $\ldots$, $p(I) = I$. Then, for each j, the Robinson–Kendall (R-K) model assumes that there exist integers $1 \leq a_j \leq I$ such that

$$\begin{aligned} \theta_{ij} &\leq \theta_{i+1\,j} \text{ for } i = 1, \ldots, a_j - 1, \\ \theta_{i+1\,j} &\leq \theta_{ij} \text{ for } i = a_j, \ldots, I-1. \end{aligned} \tag{5.11}$$

Note that when a_j is either 1 or I only one set of inequalities in the above equations is required and the other set is redundant. A matrix Θ satisfying Equation 5.11 is called a Q-matrix (for theoretical work on such matrices see, for example, Kendall 1971; Laxton 1976). In practice the true chronological order is unknown and one attempts to find an order $p(1), p(2), \ldots, p(I)$ such that Θ is a Q-matrix for a set of unknown integers $a_j, j = 1, \ldots, J$, where the matrix Θ is random and follows the posterior distribution in Equation 5.10.

The model in Equation 5.11 is overly prescriptive for most real archaeological data since the strict, temporal, unimodal sequence assumed in the R-K model may be violated because of the nature of use and discard of objects in the past and/or because of the uncertainty associated with the random nature of archaeological recovery. To account for this type of violation consider the following extension. Suppose that the matrix Φ is a Q-matrix in the natural order and let $|| \cdot ||$ denote a suitable distance measure between two matrices Θ and Φ. Here I adopt the Kullback–Leibler distance given by

$$||\Theta - \Phi|| = \sum_{ij} \theta_{ij} \log(\theta_{ij}/\phi_{ij}). \tag{5.12}$$

The extended model is then, for pre-specified $\epsilon > 0$, that there exists a matrix Θ which also satisfies the extended Robinson–Kendall model in the natural order if

$$||\Theta - \Phi|| \leq \epsilon.$$

It is clear that when ϵ is chosen to be zero the extended model reduces to the model in Equation 5.11. In this sense the parameter ϵ dictates how much relaxation I want to allow my models to have over the strict and deterministic Robinson–Kendall model. A large value of ϵ will produce all possible permutations that represent plausible seriations of the data. On the other hand, smaller values will typically produce only a few of the possible permutations of the rows for seriation.

5.5.2 Models for Correspondence Analysis

Correspondence analysis (CA) is viewed as an alternative to adopting the Robinson–Kendall model for seriation, see for example Baxter (1994, chap. 5) and Goodman (1986), but it has usually been used only in an exploratory fashion in archaeology. Following Buck and Sahu (2000) I adopt a model-based approach using hierarchical Bayesian models.

In the first stage of model building assume that $\mathbf{y}$ has a multinomial distribution with parameters N and $\boldsymbol{\theta}$ as previously, see Equation 5.9. Now assume that

$$\theta_{ij} = \theta_{i+} \, \theta_{+j} \, (1 + \lambda \, u_i \, v_j), \tag{5.13}$$

where $0 \leq \lambda \leq 1$ and u_i and v_j are unknown row and column scores satisfying the constraints

$$\sum_{i=1}^{I} u_i \, \theta_{i+} = \sum_{j=1}^{J} v_j \, \theta_{+j} = 0, \quad \sum_{i=1}^{I} u_i^2 \, \theta_{i+} = \sum_{j=1}^{J} v_j^2 \, \theta_{+j} = 1,$$

where $\theta_{i+} = \sum_{j=1}^{J} \theta_{ij}$ and $\theta_{+j} = \sum_{i=1}^{I} \theta_{ij}$. The above constraints orthogonalize and normalize the row and column scores, u_i and v_j. The parameter λ is called the canonical correlation and it is the principal eigenvalue (with the row score vector as the eigenvector) for the χ^2 distance matrix between the observed and the fitted cell counts in the contingency table. The chronological order produced by the CA is taken as the ordering of the score vector $u_1, u_2, \ldots, u_I$. Buck and Sahu (2000) detail how to specify prior distributions for the unknown parameters, λ, u_i, v_j, θ_{i+} and θ_{+j}.

5.5.3 Results

Return to the stone tools data example described in the introduction. The extended R-K model chooses the relative order $(2, 5, 3, 6, 1, 4)$ overwhelmingly while the CA model chooses the order $(3, 6, 5, 2, 1, 4)$. We can choose between the two models, hence the orders, using the Bayesian model choice methods.

I use the decision-theoretic approach of model selection to choose between the two models. The expected values of the loss function under different models are presented in Table 5.3. The extended R-K model with any value of ϵ has substantially lower expected loss values than the model for CA. Hence the extended R-K model is quite emphatically selected using this criterion.

The DIC values for the R-K model with $\epsilon = 10^{-2}$ and the model for correspondence analysis are 51.2 and 395.5, respectively, see Table 5.3. Thus the DIC also selects the extended R-K model which is simpler than the model used for correspondence analysis. By extrapolating the theoretical results obtained in Section 5.3, I intuitively conclude that the Bayes factor will also select the simpler R-K model. Although such extrapolation may not always hold, I do not recommend the calculation of the Bayes factor. The calculation is much more involved and can be numerically unstable because of the constrained nature of the parameter space under the above models. See Chapter 6 of Chen et al. (2000) for similar examples on calculation of the Bayes factor for models with constrained parameters.

Table 5.3. Model choice for the stone tools data.

Model	Expected loss	DIC
R-K ($\epsilon = 10^{-2}$)	59.5	51.2
R-K ($\epsilon = 10^{-3}$)	57.0	49.7
R-K ($\epsilon = 10^{-4}$)	55.2	47.3
CA	427.1	395.5

5.6 Discussion

In this chapter I have discussed and illustrated Bayesian model choice methods both for relative and absolute chronology building problems. Three different model choice methods have been compared and illustrated with practical examples. The chapter also points out the pressing need for adopting formal model choice methods for more complex future models which seem appropriate for archaeological data interpretation. For many purists the Bayes factor is the most appropriate tool that conveys the inferential content of the data. In this chapter, however, I have not taken such a strong view. Instead, I have presented three different competitive criteria for model choice.

There are other Bayesian model choice methods which can also be used for model comparison. For example, a Bayesian computation method known as the reversible jump MCMC (Green 1995) can be used to obtain the posterior probabilities of a number of competing models belonging to a certain structured class of models. This method is not considered because the models compared here, e.g. the extended R-K model and the model for correspondence analysis, do not belong to any class of structured nested models. Also, the two models compared in Section 5.4 correspond to two different prior distributions which cannot be written as subsets of a super-structured nested model.

The proposed Bayesian model fitting and model choice methods are attractive because these do not rely on asymptotic arguments (which usually hold for large data sets) unlike many classical methods of statistical inference, e.g. the likelihood ratio test. Such arguments are often invalid for archaeological inference problems since the associated data sets are small.

This chapter illustrates the potential of Bayesian model choice methods for statistical inference. The different Bayesian models (and prior distributions) may lead to different sets of conclusions which can be contradictory. The proposed model choice methods provide the justification for choosing one set of inferential conclusions over others.

References

Aitkin, M. (1997). The calibration of P-values, posterior Bayes factors and the AIC from the posterior distribution of the likelihood. *Statistics and Computing*, **7**, 253–261.

Baxter, M. J. (1994). *Exploratory multivariate analysis in archaeology*. Edinburgh University Press, Edinburgh.

Buck, C. E., Cavanagh, W. G. and Litton, C. D. (1996). *Bayesian approach to interpreting archaeological data*. John Wiley, Chichester.

Buck, C. E. and Sahu, S. K. (2000). Bayesian models for relative archaeological chronology building. *Applied Statistics*, **49**, 423–440.

Chen, M. H., Shao, Q. M. and Ibrahim, J. G. (2000). *Monte Carlo methods in Bayesian computation*. Wiley, New York.

DiCiccio, T. J., Kass, R. E., Raftery, A. and Wasserman, L. (1997). Computing Bayes factors by combining simulation and asymptotic approximations. *Journal of the American Statistical Association*, **92**, 903–915.

Geisser, S. and Eddy, W. (1979). A predictive approach to model selection. *Journal of the American Statistical Association*, **74**, 153–160.

Gelfand, A. E. (1996). Model determination using sampling based methods. In W. R. Gilks, S. Richardson and D. J. Spiegelhalter (eds.), *Markov chain Monte Carlo in practice*, Chapman and Hall.

Gelfand, A. E. and Ghosh, S. (1998). Model choice: a minimum posterior predictive loss approach. *Biometrika*, **85**, 1–11.

Gelfand, A. E. and Smith, A. F. M. (1990). Sampling based approaches to calculating marginal densities. *Journal of the American Statistical Association*, **85**, 398–409.

Gilks, W., Richardson, S. and Spiegelhalter, D. (eds.) (1996). *Markov chain Monte Carlo in practice*. Chapman and Hall, London.

Gómez Portugal Aguilar, D., Litton, C. D. and O'Hagan, A. (2002). Novel statistical model for a piece-wise linear radiocarbon calibration curve. *Radiocarbon*, **44**, 195–212.

Goodman, L. A. (1986). Some useful extensions of the usual correspondence analysis approach and the usual log linear model approach in the analysis of contingency tables. *International Statistical Review*, **54**, 243–309.

Green, P. J. (1995). Reversible jump Markov chain Monte Carlo computation and Bayesian model determination. *Biometrika*, **82**, 711–732.

Jacobi, R. M., Laxton, R. R. and Switsur, V. R. (1980). Seriation and dating of mesolithic sites in southern England. *Revue d'Archéometrie*, **4**, 165–173.

Kass, R. E. and Raftery, A. E. (1995). Bayes factors and model uncertainty. *Journal of the American Statistical Association*, **90**, 773–795.

Kendall, D. G. (1971). Seriation from abundance matrices. In F. R. Hodson, D. G. Kendall and P. Tautu (eds.), *Mathematics in the archaeological and historical sciences*, Edinburgh University Press, Edinburgh, 215–252.

Laud, P. W. and Ibrahim, J. G. (1995). Predictive model selection. *Journal of the Royal Statistical Society*, **57**, 247–262.

Laxton, R. R. (1976). A measure of pre-Q-ness with applications to archaeology. *Journal of Archaeological Science*, **3**, 43–54.

Nicholls, G. and Jones, M. (2001). Radiocarbon dating with temporal order constraints. *Applied Statistics*, **50**, 503–521.

Reece, R. (1994). Are Bayesian statistics useful to archaeological reasoning? *Antiquity*, **68**, 848–850.

Rubin, D. B. (1984). Bayesianly justifiable and relevant frequency calculations for the applied statistician. *Annals of Statistics*, **12**, 1151–1172.

Shennan, S. (1998). *Quantifying archaeology*. Edinburgh University Press, Edinburgh.

Spiegelhalter, D. J., Best, N. G., Carlin, B. P. and van der Linde, A. (2002). Bayesian measures of model complexity and fit (with discussion). *Journal of the Royal Statistical Society B*, **64**, 583–639.

Stuiver, M. K., Reimer, P. J. and Braziunas, T. F. (1998). High-precision radiocarbon age calibration for terrestrial and marine samples. *Radiocarbon*, **40**, 1127–1151.

6

Complicated Relations and Blind Dating: Formal Analysis of Relative Chronological Structures

Mads Kähler Holst

Summary. Relative chronology building is an essential part of chronology construction in archaeology. Whilst Chapters 1, 2, 3, 5 and 11 use relative chronological information in statistical analyses of dates assuming that it is reliable, this chapter considers relative dating evidence and the uncertainty associated with it in more detail. At most archaeological sites a formal treatment of the relative chronological evidence is first and foremost associated with stratigraphy which is represented using the Harris matrix, but the relative chronological evidence uncovered at excavations includes a much broader range of observations. These observations are often characterized by uncertainty or ambiguity as regards their temporal implications. The first part of this chapter considers different approaches in the use of relative chronology at excavations and, subsequently, a proposal is outlined for a new procedure. The procedure involves four elements: a system of description enabling the formal representation of the information, an outline of a database structure capable of handling the complexity and the dynamic properties of the data, a method of analysis, and finally some considerations on the interpretative use of the analytical results.

6.1 Introduction

Most chronology construction within archaeology directly or indirectly relies on relative chronological analysis of excavation data, giving prominence to the associated methods in the archaeological practice. The obvious example of relative chronological evidence at excavations is stratigraphy, which forms the basis of most chronological orderings of complex sites by using the succession of deposits to reconstruct a chronological sequence. Stratigraphy is, however, only one of many types of observations with a bearing on the interpretation of the relative chronology. Other examples could be the identification of a group of features such as segments of walls forming a superior architectural structure which joins the individual features by a relation of contemporaneity, or a situation where a stratum in an inlet characterized by marine fauna predates the beach ridge formation, which transformed the inlet to a fresh water environment.

With a multitude of different types of explanations, the handling of relative chronological data from excavations is not entirely unproblematic, and the characteristics of the data have proved a constant challenge. The relative nature of the data results in a high degree of complexity with data organized in a comprehensive relational system that is badly arranged from the start. The system lacks fixed points and occasionally it has rather complex relations, which depend on interpretations of a wide range of field observations. In addition, the partial and fragmented nature of the archaeological record is reflected in the relative chronological information, which often hinders a coherent representation of all information. These problems have led to the development of several different practices of relative chronological analyses.

This chapter presents a method that will allow the formal treatment of the relative chronological evidence of excavations. Section 6.2 discusses some of the general problems of relative chronological analysis through a review of the present archaeological practice. The section emphasizes the need for further development of formalized methods capable of representing and analysing the complex and ambiguous data often encountered at excavations. Section 6.3 proposes a theoretical basis for such a set of methods in the form of a descriptive system capable of representing the relative chronological information. The descriptive system is primarily intended as an attempt to permit the formalization and execution of the different practices of relative chronological analysis within a common framework of formal representation. Sections 6.4 and 6.5 deal with the analytical implementation of the system through a database application and analytical software. Finally, the interpretation of the analytical results obtained by the method is discussed in Section 6.6.

6.2 Different Approaches

Since the first appearance, in the middle of the 19th century, of relative chronology as an explicit method in archaeology, there has been a constant development of new approaches and a steady refinement of the existing methods corresponding to still growing needs for more detail and precision in the temporal statements as new archaeological problems arose and chronological schemes were refined (Gräslund 1974, 1987; Trigger 1989; Renfrew and Bahn 1991, 101ff).

This development has also applied to excavations, where the handling of relative chronological information has become a key element in defining the excavation methodology. At the same time the relative chronological information from excavations has played an essential role in the construction of chronology schemes in general (Barker 1993, 224ff). Over time several more or less well-defined methodological schools have come into existence. This is particularly obvious in connection with stratigraphy, which undoubtedly constitutes the most formalized and explicitly treated part of the analysis of excavation data, not least due to the introduction of the Harris matrix and its associated theory characterized by strict rules of inference and a rigorous selection of the usable evidence (Figure 6.1) (Harris 1975, 1989; Harris et al. 1993). In principle, the Harris matrix is a graph representation, in which each individual deposition episode is depicted as a node in the graph, and the relative chronological relations are shown as edges between the nodes. The nodes are events; they function as points in time rather than time spans. As regards the relations the principal options are that either (a) something is later than or earlier than something else or (b) there are no relations between the two. The possibility of representing the situation where two apparently independent entities originally were part of a whole has, however, also been added (Harris 1989, 34ff). Finally, in some variations of the approach contemporaneity is included as an option (Herzog 1993).

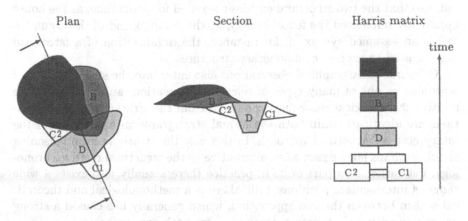

Fig. 6.1. Examples of representations of a stratigraphical sequence in plan, in section and in the Harris matrix.

At the same time as we have the formal treatment of the Harris matrix approach, relative chronological information is, however, also used in a much less formal and often implicit way, mainly during the excavation in the continuous, dialectic interpretation of the site, but to some extent also in the post-excavation analyses (Hodder 1997, 1999). Here relative chronology is used to establish and test working hypotheses on the organization and development of the site. This approach involves a more complex interpretation of observations and a wider range of possible temporal relations compared with the formal stratigraphical analysis (Hvass 1979, 1985).

The phenomenon is especially obvious at excavations with abundant architectural remains and activity areas, which in archaeology is often grouped under the term structures. An example of such a site is the third to seventh century AD village of Nørre Snede in Central Jutland, Denmark, depicted in

Figure 6.2. The accumulated traces of all buildings constructed through the 500-year time-span of the settlement are found complexly intertwined within a relatively limited area. It is immediately obvious from the confusion of structures that any analyses of the settlement structure presuppose a recovery of the relative chronological order of the buildings. At excavations of this type of site a sort of "logic", or perhaps more accurately a reasoning, of spatial behaviour and organization is used to suggest, for instance, contemporaneity, non-contemporaneity, and continuity (Sharon 1995b, 752f; Holst 1999). The information used in this line of reasoning ranges from simple observations of connections between structures to complex interpretations of large data sets. An example from the simple end of the continuum could be the blocking of an entrance in a house by another structure, which would indicate that the two structures were not contemporary. Somewhat more complex is the identification of a fence, whose course is diverted around the traces of a house. Temporally this could imply that the house was founded before the fence, but also that the two structural entities coexisted for some time, as the house apparently influenced the fence. Finally, at the complex end of the argumentation an assumed syntax of, for instance, the organization of a farmstead may be used to suggest contemporary structures.

Of course stratigraphical observations also enter into the structural line of reasoning as one of many types of relevant information, and principally the relative chronological consequences derived from the stratigraphical observations are identical within both the formal stratigraphical approach and the interpretative, structural approach. In this way the stratigraphical reasoning should perhaps not be seen as an alternative to the structural relative chronology, but rather as a part of it. In practice there actually also exists a wide range of intermediate positions. Still, there is a methodological and theoretical schism between the two approaches, which generally has caused a strong opposition with unfortunate consequences for both approaches.

The general lack of formalization of the structural approach combined with the often rather complex temporal implications of the observations used and the inclusion of uncertain and ambiguous information frequently lead to major problems of both controlling and presenting the logical consistency of the relative chronological arguments on anything but very small data sets (Holst 2001). Furthermore, it is difficult to evaluate the chosen solution both in relation to alternative solutions, which inevitably will exist with the inclusion of the uncertain and the strongly interpretatively based information, and in relation to the different observations used and the concept of their temporal implications. How would the exclusion of certain observations and different concepts of relative chronological consequences influence the interpretation? We are in this way presented with a relative chronological result, but the process of achieving the result is shrouded in darkness. In reality this kind of blind dating leaves us with the option of either accepting the result and its preconditions in its entirety or rejecting the result completely (Madsen 1995).

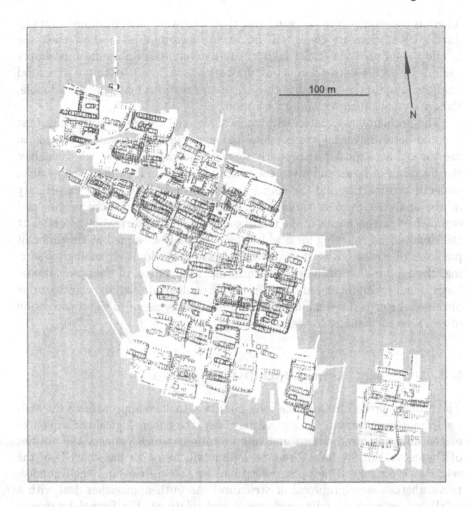

Fig. 6.2. Excavation plan of the settlement of Nørre Snede, Central Jutland, Denmark. The site is an example of an excavation with a large number of structures accumulated within the same area over a long time span.

Formal stratigraphic analysis, on the other hand, is easily controllable and capable of handling very large data sets, not least due to the possibility of computer-aided analysis. The problems here lie in the limited flexibility of the method. It demands unambiguous and certain information within a discipline where those exact properties are very much a rarity due to the fragmented and partial nature of its source material. Consequently, the method may achieve a strict order based on a rigid arrangement of relationships, but at the same time it sets standards on the generation of the data, and these standards are not always quite in concordance with the character of the source material.

There is consequently a potential risk that somewhat uncertain observations during excavation are converted into secure information to allow their use in the formal analysis (Holst 2001). Alternatively, the information is simply excluded together with a number of types of observations used in the structural analysis, despite the possibility that these observations may contain valuable, though somewhat uncertain, information.

The problems associated with each of the two approaches indicate that it could be an advantage to combine the strength of the two. To be able to use the archaeological data properly we need the formalization, which enables the handling of large and often complex data sets, and we need the flexible, dialectic possibilities and the ability to deal with ambiguity and uncertainty in the analyses. Such a combination could contribute to more precise and complete representations of the relative chronological information available at the excavations, allowing for a wider range of analyses aimed at the different purposes of the use of relative chronology at excavations: chronological sorting, the explanation of the site organization, and control of the observational basis. Finally, the increased detail in the chronological analysis which may be obtained through the combination could potentially assist in the construction of more detailed chronological schemes.

6.3 Representation: Descriptive System

The primary precondition of the combination of the two approaches is the ability to represent formally in a consistent way, the chronological consequences of the different observations and interpretations which connect the entities of the excavation. A formal representation will allow formal analysis of the relative chronological network constituted by the entities and their connections whether stratigraphical or structural. As both approaches deal with a relational network of relative chronological relations, the formal treatment adopted in the Harris matrix with its graph representation forms an evident starting point.

The chronological consequences used in the structural analysis, on the other hand, are generally more complex and diverse. One basic idea is that the founding of a structure is influenced by pre-existing structures in agreement with rational principles of spatial organization, as in the example of the structure blocking the entrance. The observations used in this way refer to both the starting point and the life span of the structures, which is a fundamental difference from the exclusive "points in time" concept of the Harris matrix. Also, different ideas of formation processes on archaeological sites are occasionally used to derive chronological consequences, and these may refer to both the founding and the demolition of the structures. For instance, the remains of burned debris in the fill of a posthole belonging to one structure in the vicinity of another structure destroyed by fire may be seen as an indication that the posthole post-dates the destruction of the burned structure. Finally, there are observations which indicate that entities did not exist simultaneously, such as overlapping houses without stratigraphical relations and more ambiguous expressions as indications of continuity without certain information on the temporal order of the two. It should be obvious from these examples that the range of possible chronological consequences in the structural analysis of relative chronology is wide.

The development of a formal representation of the temporal interpretation comprises two tasks: firstly, a representation of the archaeological entities, which means both the deposits and the structures, and secondly, a representation of the relations between the entities.

The extended demands of the structural approach can be conceived as the result of a concept of the structures as covering a time span delimited by the founding and the demolition of the structure. Both the founding and the demolition appear in the relative chronological argumentation as points in time, more specifically the points where the structures respectively began and ceased potential influences on other structures. This is of course an approximation of a somewhat more complex past reality, but it is generally an acceptable approximation which does not cause logical problems, and it certainly reflects the normal practice within relative chronological analyses of both stratigraphical and structural excavation data. In the few situations where the approximation proves a problem, a different expression of the temporal implications will be an efficient and precise solution. This approach will be developed further below.

The chronological consequences in some of the examples above also included references to the life span of the structures in general and not the start and end points of the structures. However, as the life span can also be expressed by using simply the start and end points of the structure, it is not necessary to introduce the life span as an independent entity in the formal descriptive system.

In a graphical representation, the outlined concept of the structures corresponds to each structure being represented by two nodes, the start and the end. These nodes normally function as points in time that are related by an earlier than–later than relation. The formal properties of the nodes are in this way principally identical to the properties of the nodes of the Harris matrix. The difference lies in the conceptual tie to the archaeological entities.

Of course, deposits in a stratigraphical sequence can, in the same way as structures, be represented by two nodes indicating a start and an end of the deposition. This emphasizes a perception of the deposition as an occurrence of some duration. Alternatively the momentary event concept of the Harris matrix can be maintained.

Having established a way of formally representing the archaeological entities, the next step concerns the definition of the relations between the entities.

The connecting observations and interpretations of the excavation data imply chronological consequences, which can be expressed as relations between the start and the end of each structure (or deposit). These relations are purely relative chronological; that is, the nodes are connected with the well-known relations earlier than–later than and contemporary with. However, some extensions are necessary to represent the ambiguity and uncertainty of the archaeological data (Holst 1999). Normally, we wish all relations to be true, which means logically the relations are combined with an AND connective. By introducing an either-or connective (XOR), ambiguity and uncertainty can be included. It is thus possible to describe, for instance, a situation where two structures cannot have existed simultaneously as EITHER *the start of structure 1 is later than or contemporary with the end of structure 2* OR *the end of structure 1 is earlier than or contemporary with the start of structure 2.* The resulting expression is obviously somewhat difficult to embrace, and a clear notation is necessary. The following symbols may represent the different relations and connectives, including symbols for the expressions *later than or contemporary with* and *earlier than or contemporary with.*

$=$	for the relation	*contemporary with*
$<$	for the relation	*earlier than*
$>$	for the relation	*later than*
$\leq$	for the relation	*earlier than or contemporary with*
$\geq$	for the relation	*later than or contemporary with*
AND	for the connective	*and*
XOR	for the connective	*either... or*

The example of the two non-contemporary structures, can in this way be described as

$$\text{Structure1(start)} \geq \text{Structure2(end)}$$
$$\text{XOR}$$
$$\text{Structure1(end)} \leq \text{Structure2(start)}$$

As mentioned above, the assumption of a temporally clearly defined start and end of the structures may occasionally be problematic. In these instances it can be an advantage to operate with a concept of broad contemporaneity, which has been introduced in stratigraphical analysis in an attempt to deal with some of the problems caused by time spans (Sharon 1995b, 753ff). The broad contemporaneity allows for some degree of flexibility compared with the strict contemporaneity, which locks the entities completely in relation to each other. The introduction of broad contemporaneity leads to three new relations:

$\equiv$	for the relation	*broadly contemporary*
$\lesssim$	for the relation	*earlier than or broadly contemporary*
$\gtrsim$	for the relation	*later than or broadly contemporary*

The descriptive principles can be seen as a formal language enabling expression of the different chronological connections possible between two structures (Figure 6.3). An example based upon the concept of a momentary start and end of the structures is presented in Figure 6.4. The list is not exhaustive, and the range of possibilities is obviously wide, allowing quite detailed relative chronological statements. For instance, it is possible to distinguish between different types of contemporaneity with varying restrictions on the relative positions of the structures, and an assumed completeness of a continuous sequence of structures would lead to a different set of relations than a sequence where the continuity cannot be assumed.

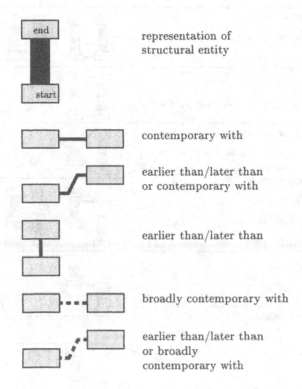

Fig. 6.3. The basic principles of graphical representation of the structural entities and the relations between the nodes.

6.4 Data Handling: Adaptation and Flexibility

Having established some general principles of description the next step is the implementation of these principles in the analyses of data. An important part of this task is to ensure the capabilities of handling data, which implies the de-

Temporal consequences   *Graphical representation*

Partial asynchronism

a) $X(start) < Y(start)$ **XOR** $X(start) > Y(start)$

Full asynchronism

b) $X(start) \geqq Y(end)$ **XOR** $X(end) \leqq Y(start)$

General synchronism

c) $X(start) < Y(end)$ **AND** $X(end) > Y(start)$

Specific synchronism

d) $X(start) = Y(start)$

e) $X(end) = Y(end)$

Full synchronism

f) $X(start) = Y(start)$ **AND** $X(end) = Y(end)$

Asymmetrical synchronism

g) $X(start) \leqq Y(start)$ **AND** $X(end) \geqq Y(end)$

General diachronism

h) $X(end) < Y(end)$

i) $X(start) < Y(start)$

j) $X(start) < Y(end)$

Full diachronism

k) $X(end) \leqq Y(start)$

General continuity

l) $X(end) = Y(start)$ **XOR** $X(start) = Y(end)$

Directional continuity

m) $X(start) = Y(end)$

General discontinuity

n) $X(end) < Y(start)$ **XOR** $X(start) > Y(end)$

Directional discontinuity

o) $X(start) > Y(end)$

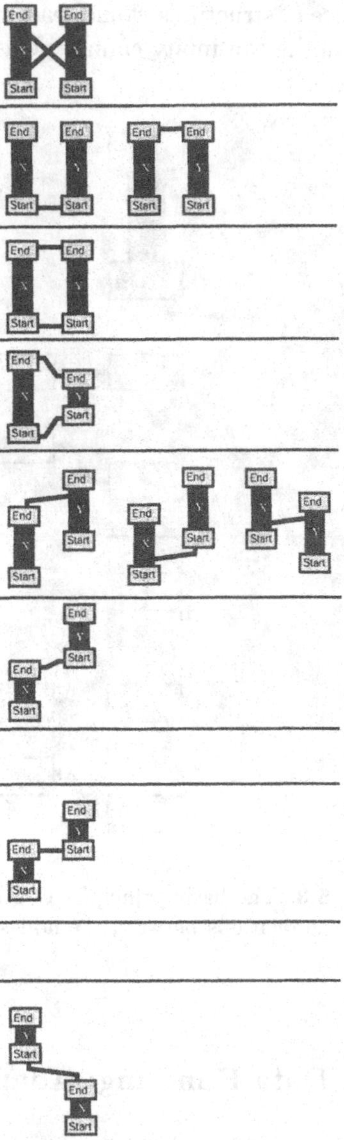

Fig. 6.4. An example of a classification and definition of chronological consequences.

velopment of a database with an incorporation of the principles of description outlined above.

A major disagreement between the different approaches to relative chronology obviously lies in what observations and interpretations of observations to include in the analysis. The formal stratigraphical approach is by far the most restrictive, whereas some structural approaches appear almost all embracing. This discrepancy reflects different critical valuations, which is fully legitimate, as there may be different demands on certainty in relation to explanation. Thus, the strict, formal analysis of stratigraphy represents the emphasis on certainty and is well suited for the study of traditional chronological problems, where it delivers a solid basis for a relative temporal ordering of, for instance, artefacts contained in the strata. The structural approach on, the other hand, is clearly associated with attempts to reconstruct and explain the site formation, where for a long time it seems to have proved itself a useful interpretative tool with or without formalization.

Another reason for variations in the indications of connections that are acceptable in the relative chronological analyses can be found in the fact that some of the data of the structurally based analysis of relative chronology presupposes assumptions on prehistoric behavioural patterns, and insofar as the chronological analysis is intended to elucidate exactly prehistoric behaviour, these data evidently have to be excluded from the chronological analysis in order to prevent circularity.

For partly the same reasons, often there are also discrepancies in the chronological consequences of the different connecting data, as the exact formulation of the consequences is dependent on various preconditions, some of which may not be acceptable in specific analysis. This leads to modifications of the chronological consequences. Similarly, different research traditions and schools tend to have varying concepts of the consequences of the same observations, and, finally, different research problems may require different details, which also leads to different definitions of the chronological consequences.

Altogether these characteristics demonstrate the basic need for the flexible inclusion and exclusion of both (a) the different observations connecting the entities and (b) a dynamic combination of the observations with different types of chronological consequences dependent on research problems and traditions. They also indicate that it can be difficult to maintain the existence of only one correct relative chronological sorting of any given set of entities. This reflects the idea that basically the relative chronological models we create at excavations are not primarily models of past sequences of events, but rather models of our interpretation of the surviving indications of that past, and they are far from being unequivocal.

The demand of flexibility is basically a question of data handling and reflects the need for a relatively complex database design. The very problems of handling the complexity of excavation data have been in focus in connection with the development of a generalized database system in archaeology, where the use of relational database principles has been advocated as a possible solution (Madsen 1999, 2001). The demands of flexibility in the combination and use of chronological consequences and observations in the relative chronological analysis can actually also be met with a comparatively compact relational database design, as represented in outline by Figure 6.5.

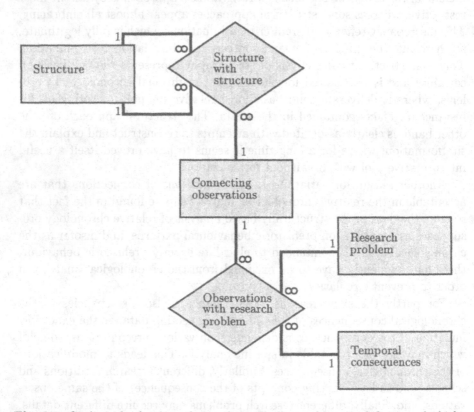

Fig. 6.5. Outline of a database design that will allow a dynamic definition of which observations are legitimate within different research problems and which chronological consequences the individual observations imply.

The primary components are the structures identified at the excavation and the observations or interpretations connecting them. The individual structures and the different types of connecting observations are defined in separate tables. The table *structure with structure* permits the interrelating of the structures by the different connecting observations. These three tables of the database can be considered a sort of compilation of the basic data closely associated with the archaeological source material as encountered at the excavation, and as such they possess a certain integrity, notwithstanding the rather complex interpretations which are occasionally involved (Gardin 1980; Madsen 1995).

The remaining part of the database is the explanatory part and contains two points of entry. The different types of chronological consequences can be defined in one table, whereas another is reserved for research problems. Through the table entitled *observations with research problem* it is possible to define which observations are usable within each research problem and which type of chronological consequences the specific observation implies given the current research problem. Occasionally there may also be disagreement about which types of entities to include in an analysis. This problem could obviously be solved in a similar way.

In this way the database allows the definition of different types of connecting observations and different types of chronological consequences. At the same time it is also possible to define a series of research problems; within each research problem the legitimate types of connecting observations can be selected and a specific linking of the individual observations and specific temporal implications can be defined. By including a specific research problem in the queries in the database only the acceptable connecting observations within the research problem referred to and the matching chronological consequences will be returned. Queries on different problems will return different data sets, but insofar as the same principles of formal representation are used to describe all temporal implications, the analytical process will remain the same, and the analytical results can be compared immediately. The database thus delivers the desired flexibility. In addition a major advantage of this type of recording is the explicit formulation of the principles of acceptable inference within each research problem.

6.5 Analysis

Having established a system of formal representation of the relative chronological relations between the entities of the excavation and implemented this system in the database of the excavation, the next step is the analysis of the relational network.

The choice of analysis reflects a combination of the available data, which are defined by the formal representation, and the purpose of the analysis. As regards the purpose, the relative chronological sorting is of course the overall objective, but at the same time the explanation of the site formation and the check of the consistency of the observations are secondary objectives, with somewhat different requirements.

As with the Harris matrix approach we are still basically dealing with series of related entities, that is, a relational network. However, the range of possible relations has been extended, exclusive disjunctions have been introduced and in the analysis we include relations of varying certainty and quality. The introduction of the disjunctions is of fundamental importance, as there is no longer necessarily only one unambiguous, logically consistent solution, but rather a wide range of possible solutions. This reflects the ambiguity and the fragmented and partial nature of the archaeological record. Furthermore, due to the inclusion of the uncertain relations, we must also allow for some degree of logical inconsistency in the solutions. Even though a sorting does not concur perfectly with the chronological consequences of all observations, it may still be relevant, and the objective of the relative chronological analysis has changed from a search for the correct solution to a dialectic exploration of different optimal and near optimal sortings. This may seem an unsatisfying position, but it must be stressed that it is basically a reflection of the nature of the archaeological data.

Mathematically, the modification of the system of description also transforms the problem of sorting considerably. The Harris matrix description with, in principle, only one asymmetrical type of relation results in a relatively simple directed web, which can be sorted with topological algorithms. With the introduction of the disjunctions we are faced with a non-deterministic polynomial complete problem (a so-called NP-complete problem). The obvious way to handle this type of problem is the use of heuristic optimization algorithms, however, in connection with the interpretation of results it is important to emphasize that with the exception of the perfect solutions where all conditions are true, there is no guarantee that a specific sorting represents the absolute optimal solution. The heuristic approach has already been implemented in stratigraphical analysis by Sharon, who used a simulated annealing algorithm to enable a formal analysis with a more flexible interpretation of stratigraphical sequences compared with the Harris matrix approach, which allows among other things the search for alternative solutions and the use of the concept of broad contemporaneity (Sharon 1995a,b).

The simulated annealing algorithm belongs to the generalized heuristic optimization algorithms, and can be modified to handle both the disjunctions and the extended range of possible relations introduced in the system of description presented above. A computer program named *Tempo* has been developed based on the *POSAR* program developed by Sharon, but with modifications to accommodate the descriptive principles presented above (obtainable at http://www.math.ku.dk/~holst/tempo). It can perform a simulated annealing on data sets generated according to the principles of the descriptive system, and it presents a graph representation of the sorting with both the nodes and the explicit relations depicted. Relations inconsistent with the sorting are emphasized (Figure 6.6). The graph is interactive in the sense that the nodes can be moved manually in a continuous check of its logical consistency, whereby, for instance, the tolerance of the sorting can be tested and the possibility of specific modifications can be studied. Repeated runs of the algorithm either as a continuance of an initial sorting or on an entirely different random initial order of the nodes may also reveal several alternative solutions, which may prove quite different depending on the data. In addition, the algorithm contains a number of adjustable parameters, which also influence the sorting and can be used exploratively.

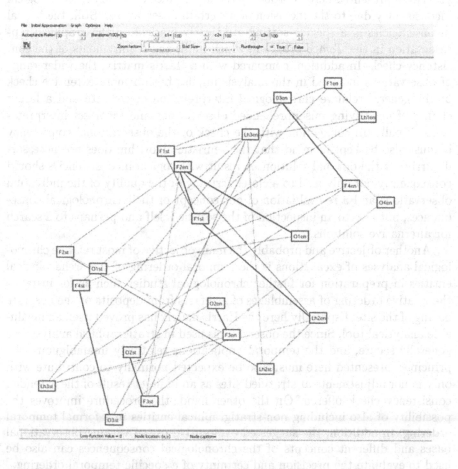

Fig. 6.6. Example of sorted entities in the *Tempo* program. The time axis runs from the bottom to the top. The different grey shades of the lines indicate different types of relations and discrepancies between the logical relation and the actual orientation in the graph are emphasized with a thicker line.

6.6 Interpretation

A major reason for the different approaches to relative chronological analysis and the ensuing demand of analytical flexibility lies in the varying purposes of the relative chronological analysis. Therefore, in the following attention will be directed at the different interpretative possibilities of the relative chronological analysis in relation to the procedure presented above.

A basic and initial task of relative chronology at excavations is the control of consistency of the observations and interpretations of the excavation. Compared with traditional practices, the major conceptual difference of the procedure presented here is probably the acceptance of some degree of logical inconsistency due to the inclusion of uncertain observations. Still, the logical inconsistencies of a specific sorting are clearly marked on the graphical representation in the *Tempo* program, which fulfils the requirements of the consistency check. In addition, compared with a Harris matrix, the wider range of observations included in the analysis implies both a more extensive check of the general relative chronological interpretation of the site and a larger chance of identifying misapprehended observations and incorrect interpretations. Finally, in connection with the check of the observational consistency it must also be kept in mind that the heuristic algorithm does not necessarily arrive at the optimal solution. A result with logical inconsistencies should consequently not only lead to a closer scrutiny of the quality of the individual observations and a re-evaluation of our concept of their chronological consequences, but also to an inspection of the sorting itself and perhaps to a search for alternative solutions.

Another objective and probably the main objective of most relative chronological analyses of excavations is the temporal ordering of the archaeological entities in preparation for further chronological studies such as for instance the relative ordering of assemblages of artefacts from deposits or the absolute dating of the site. Especially here the Harris matrix has proved itself an invaluable analytical tool. Since the observations used in stratigraphical analysis are generally secure, and the temporal implications relatively unambiguous, the principles presented here must also be expected normally to contribute with only minor adjustments at stratified sites as an indirect result of the extended consistency check offered. On the other hand, the procedure improves the possibility of also including non-stratigraphical entities in a formal temporal ordering. In addition, the suggested explorative use of varying observational bases and different concepts of the chronological consequences can also be used to evaluate the precision and certainty of a specific temporal ordering.

The explanation of site organization and formation, which is the substantiated attempt to reconstruct the spatial organization of a site over time, also relies heavily on relative chronological information, although the objective here is not chronological in a strict sense. The structural approach is traditionally closely associated with the explanation of site formation, and as the procedure presented here to a large extent is an attempt to formalize the structural approach, the main potentials of the procedure are also associated with the explanation of site formation. The main advantage partly lies in the possibility of including the wide range of observations and interpretations necessary to propose a theoretical model of the organization and its development and partly in exactly the formalization that enables a handling of the otherwise unmanageable complexity of the information. The procedure in this way contributes and is closely related to the relatively few other formalized approaches to structural analysis (Simmons et al. 1993; Sharon 1995a,b).

The final potential of the relative chronological information from excavations is related to the explanation of site formation, but is rarely exploited.

It can be termed the study of temporal structure, which comprises the analysis of how changes occur in settlements, such as for instance examinations of whether alterations occur contemporaneously or gradually. The main prerequisite of this type of analysis is an ability to describe the temporal relations between the entities in great detail, which is precisely what is made possible by representing each entity with two points: start and end nodes (Holst 1999).

6.7 Conclusion

The analytical potential of the relative chronological information derived from excavations is obviously vast, and the relative chronological analyses of excavation data will probably remain the principal basis for handling chronological and temporal problems within sites in spite of the growing possibilities and the refinement of absolute dating. A large unused potential still remains in the relative chronological information, and it is possible to reach a surprising level of detail in the temporal resolution of sites with complex stratigraphy or structural remains. However, the more detailed we wish our picture of the past to be, the stronger the demand for a formal treatment becomes.

The procedure of handling the relative chronological information and structures of excavations presented in this chapter can be seen as an attempt to accommodate different approaches to and practices of excavation analyses. Compared with the existing formal approaches, a much wider range of observations and other indications of relative chronological relations can be included in the formal treatment, and there is increased detail in the expression of the chronological consequences of the different observations.

When we use the extended range of observations and try to reach the higher level of detail, it is, however, also obvious that the interpretative element of the relative chronological treatment increases considerably and the relative chronological analysis becomes intimately linked with the explanation and understanding of the site formation and organization in general. Beyond the restrictions of the research problems in question, the degree of interpretation that is acceptable in different analyses must be an individual decision, and the borderline will undoubtedly vary considerably. Consequently, to comply with the different analytical purposes and personal preferences we must expect excavation analysis to often involve multiple models of the relative chronological structure of the same site, demonstrating the fact that in relative chronological analysis nothing is fixed.

Acknowledgements

I would like to thank sincerely Ilan Sharon for placing the *POSAR* program and his inspiring Ph.D. thesis at my disposal. Furthermore, I also thank Klaus

Holst for his efforts in developing the *Tempo* application and our many discussions on the analytical possibilities. Finally, thanks go to Torsten Madsen for comments on the chapter.

References

Barker, P. (1993). *Techniques of archaeological excavation*. Batsford, London, third edn.

Gardin, J.-C. (1980). *Archaeological constructs: an aspect of theoretical archaeology*. Cambridge University Press, Cambridge.

Gräslund, B. (1974). *Relativ datering: Om kronlogisk metod i nordisk arkeologi*. Bibliotek "TOR". Inst. f. arkeologi, särskilt nordeuropeisk, Uppsala University, Uppsala.

Gräslund, B. (1987). *The birth of prehistoric chronology: dating methods and dating systems in nineteenth-century Scandinavian archaeology*. Cambridge University Press, Cambridge.

Harris, E. C. (1975). The stratigraphic sequence: a question of time. *World Archaeology*, **7**, 109–121.

Harris, E. C. (1989). *Principles of archaeological stratigraphy*. Academic Press, London, second edn.

Harris, E. C., Brown, M. R., III and Brown, G. J. (eds.) (1993). *Practices of archaeological stratigraphy*. Academic Press, London.

Herzog, I. (1993). Computer-aided Harris matrix generation. In E. C. Harris, M. R. Brown, III and G. J. Brown (eds.), *Practices of archaeological stratigraphy*, Academic Press, London, 201–217.

Hodder, I. (1997). 'Always momentary, fluid and flexible': towards a reflexive excavation methodology. *Antiquity*, **71**, 691–700.

Hodder, I. (1999). *The archaeological process: an introduction*. Blackwell, Oxford.

Holst, M. K. (1999). The dynamic of the Iron-age village: a technique for the relative-chronological analysis of area-excavated Iron-age settlements. *Journal of Danish Archaeology*, **13**, 95–119.

Holst, M. K. (2001). Formalizing fact and fiction in four dimensions: a relational description of temporal structures in settlements. In Z. Stancic and T. Veljanovski (eds.), *Computing archaeology for understanding the past*. British Archaeological Reports, Oxford, International Series, **S931**, 159–163.

Hvass, S. (1979). Die völkerwanderungszeitliche Siedlung Vorbasse, Mitteljütland. *Acta Archaeologica*, **49**, 61–111.

Hvass, S. (1985). *Hodde, et vestjysk landsbysamfund fra ældre jernalder*. Arkæologiske Studier, vol. VII. Akademisk Forlag, Copenhagen.

Madsen, T. (1995). Archaeology between facts and fiction: the need for an explicit methodology. In M. Kuna and N. Venclová (eds.), *Whither ar-*

chaeology: papers in honour of Evzen Neustupny, Institute of Archaeology, Praha, 125–144.

Madsen, T. (1999). Coping with complexity: towards a formalised methodology of contextual archaeology. *Archeologia e Calcolatori*, **10**, 125–144.

Madsen, T. (2001). Transforming diversity into uniformity: experiments with meta-structures for database recording. In Z. Stancic and T. Veljanovski (eds.), *Computing archaeology for understanding the past*. British Archaeological Reports, Oxford, International Series, **S931**, 101–105.

Renfrew, C. and Bahn, P. (1991). *Archaeology: theories methods and practice*. Thames and Hudson, London.

Sharon, I. (1995a). *Models for stratigraphic analysis of tell sites*. Ph.D. thesis, Hebrew University, Israel.

Sharon, I. (1995b). Partial order scalogram analysis of relations – a mathematical approach to the analysis of stratigraphy. *Journal of Archaeological Science*, **22**, 751–767.

Simmons, D. M., Stachiw, M. O. and Worrell, J. E. (1993). The total site matrix: strata and structure at the Bixby Site. In E. C. Harris, M. R. Brown, III and G. J. Brown (eds.), *Practices of archaeological stratigraphy*, Academic Press, London, 181–197.

Trigger, B. G. (1989). *A history of archaeological thought*. Cambridge University Press, Cambridge.

Genealogies from Time-Stamped Sequence Data

Alexei Drummond, Geoff K. Nicholls, Allen G. Rodrigo, and Wiremu
Solomon

Summary. This chapter focuses on on-going research into chronology building tools based on genetic data from DNA sequences. By combining genetic information and radiocarbon data from fossil remains it is possible to recover genealogical structures, population size information, mutation rates and, hence, approximate chronologies for genetic trees. Since this is such a new area of research, this chapter provides consideration of the problems that need tackling, and makes a range of suggestions for modelling aspects of them. Detailed explanations of the proposed models are given and insight into the nature and size of the uncertainties associated with the chronological estimates is obtained. An illustrated case study indicates the type of problem that can already be tackled, and shows that uncertainty derives mainly from the genetic model and not the radiocarbon dates.

7.1 Introduction

The remarkable thing about using genetic material to date events, is that it is possible at all. There are two clocks in the story, one ticks at the mutation rate of DNA, the other at the coalescent rate of ancestral lineages. We know the rate constant for neither clock. We seem to be pulling ourselves up by our bootstraps, obtaining rates and dates from the one data set. It turns out that in fact we cannot date without some prior knowledge of the likely value of the date we wish to estimate. However, only very limited information is needed, as we show. In this chapter we review recently published statistical methodology for genetic dating. Our explicit treatment of the uncertainty arising from imprecisely dated sequences (Section 7.6) is new, as is our characterization of the limitations of genetic dating (in Section 7.4).

What are we dating here? We can estimate an age for the common ancestor of the individuals whose DNA makes up the data. The amount of divergence between the DNA sequences of the sample individuals is a measure of that age. We can also estimate the time at which an individual lived, if we have an undated DNA sequence from the individual, along with dated sequences from other individuals in the same population. In Drummond et al. (2002), the authors date modern-day events, using HIV sequence data, with ages known to the day. Where radiocarbon dating is used to date fossil DNA, as in Lambert et al. (2002), analysis to date has been conditioned on point estimates of calibrated dates. This approximation is often acceptable, as we explain in Section 7.6.

What archaeological questions can we answer? We can provide estimates of dated genealogies, and in some simple cases, an estimate of the way the total population size has varied as a function of time. It is possible that direct genetic dating of fossil DNA sequences (see Section 7.8 and Figure 7.6) may in the future be used to make very crude temporal classifications, along the lines of "ancient" or "modern". However, care must be taken when gene-derived dates are used to infer cultural history. Where this link is needed, an explicit case must be made. The argument is typically based on archaeological context, but may follow from statistical considerations, as in Penny et al. (1993), who discuss the parallel genealogies of human languages and human genes. The genealogies of domestic animals and human and animal diseases may be of independent interest, whilst recent contact between human communities may potentially be resolved by the viruses shared by those communities, since viruses evolve at a much higher rate than their hosts. In Matisoo-Smith et al. (1998), the genealogy of Polynesian rat mtDNA reveals prehistoric patterns of human mobility in East Polynesia. Underhill et al. (2001) develop a model of human contact in the wider Pacific from Y-chromosome data for Pacific peoples. Before sequence data were available, ancestral inference, in archaeology as in biology, was based on higher-level trait-based data. Ammerman and Cavalli-Sforza (1984) consider the mesolithic-neolithic transition in Europe in the light of blood type and other trait-based data. The citations in this paragraph reflect the authors' personal interests.

We focus on recent developments in statistical methodology. Data-analytical tools, developed in the last decade, allow us to put error bars on the dates and rates reconstructed from sequence data. This is not trivial, as we don't usually know how the sample specimens are related. Kuhner et al. (1995) used MCMC to average over genealogies and obtain a maximum-likelihood estimate for one of the two rate parameters of the problem, if the other is known. Model averaging of this kind is computationally demanding, but statistically robust. Kuhner et al. (1995) suppose all data sequences have equal age. Importance sampling methods are used to get estimates at parameter values which were not simulated. This proves to be a weakness of the method, as the importance weights can have high variance. The Bayesian analysis given below uses MCMC to average genealogies and rate parameters simultaneously, from sequence data gathered at different times. Whilst MCMC methods have their own weaknesses, and must be used with care, they have certainly extended the range of reliable population-genetic inference. As we explain below, the output of our simulations may be used to form maximum-likelihood estimates, if so desired.

The methods described below were applied in Drummond et al. (2002), to estimate the parameters of an HIV population, and in Lambert et al. (2002), to estimate the age of the common ancestor of a collection of fossil Adelie penguin bones. In earlier work, Rambaut (2000) starts with a maximum-likelihood phylogeny with time-stamped sequence data at the leaves and estimates the mutation rate. This kind of analysis, which dates back to Felsenstein (1981), is not robust, but is convenient for exploratory work. Barnes et al. (2002) estimate the genealogy of time-stamped fossil bear sequences via parsimony.

Bayesian inference is relatively new to population genetics. Wilson and Balding (1998) and Beaumont (1999) are early examples, treating microsatellite data. Wilson et al. (2003) overlaps with Drummond et al. (2002). The natural prior on genealogies, the coalescent process of Kingman (1982a), has a frequentist interpretation as a part of the observation model. This is because the coalescent process is derived from an explicit and plausible model of inheritance structure, the Wright–Fisher model. The coalescent has cornered the market for population genetic models. However, the Wright–Fisher model applies to a single panmictic population, and is therefore inappropriate for genealogies which unite distinct species. When we survey the literature we see a variety of phylogenetic tree priors in use. The knowledge they express is more obviously subjective than is the case in the population genetic setting. Perhaps for that reason, Bayesian methods are more common in phylogenetics than in population genetics. See for example Suchard et al. (2001) and references therein.

In this chapter we lay out the methodology in the context of a model of asexual reproduction. The cells of sexually reproducing creatures contain mitochondrial organelles, which behave a bit like cells within cells. Mitochondria carry their own DNA, and that mtDNA reproduces asexually, following the maternal line. It follows that the methodology we lay out below can be used to date events in the maternal ancestral tree of creatures which reproduce sexually.

7.2 The Mutation-Clock

The mutation model described in this section is the (standard) independent neutral finite-sites mutation model of Felsenstein (1981).

Consider a DNA sequence $B_1, B_2, \ldots, B_L$ of L sites or *loci*. The character B_s at site s records the nucleotide type at that site. There are four types, the two purine bases with labels **A** and **G** and the two pyrimidine bases, with labels **C** and **T**. Let $\mathcal{C} = \{\mathbf{A}, \mathbf{C}, \mathbf{G}, \mathbf{T}\}$. At a generation the DNA sequence is copied, site by site, to make a new sequence $Y_1, Y_2, \ldots, Y_L$. In the absence of copy errors, $B_s = Y_s$ for each $s = 1, 2, \ldots, L$. However, at each copy event, there is a small chance of a copy error occurring. We will assume these errors are independent but identically distributed from one site to another.

Focus on site s, and suppose $B_s = X$ and $Y_s = Y$ with X and Y characters in $\mathcal{C}$. What is the probability that X mutates to become Y at site s in the course of a generation? A single generation is a very short time compared with the timescale at which we want to work, so we will write this probability in terms of a dimensionless relative-rate matrix, $Q_{X,Y}$, a mutation rate parameter μ, with units *mutations per year*, and ρ, the number of years per generation. If $X \neq Y$, the probability for the event is $\mu Q_{X,Y}\rho$. Consider what happens over many generations. Let B denote an ancestral sequence and B' a descendant sequence, and suppose the two sequences are separated by an interval of time t much larger than ρ. If we can ignore terms of order ρ/t, the probability $\Pr\{B'_s = Y | B_s = X\}$ to get a Y at site s in B', given there was an X at site s in B is

$$\Pr\{B'_s = Y | B_s = X\} = [\exp(\mu Q t)]_{X,Y}.$$

The exponential function of the 4×4 matrix Q is defined by the exponential series $\exp(M) = 1 + M + MM/2! + MMM/3! + \ldots$, where 1 is the 4×4 identity, and MM is just matrix multiplication. Notice that when t gets small we can approximate $\Pr\{B'_s = Y | B_s = X\}$ by the matrix $1 + \mu Q t$. The off-diagonal elements are just what we started with, $\mu Q_{X,Y}t$. The row sums of $1 + \mu Q t$ should be one (because it is a transition probability), so we need $Q_{X,X} = -\sum_{Y \neq X} Q_{X,Y}$ for the diagonal elements of Q.

In the following we will suppose, without further discussion, that all the entries in Q are known. In fact we can estimate those relative rates from the data along with all the other unknowns treated here. When the absolute rate μ is high, there are plenty of mutations, so the data pins down relative rates fairly tightly. Drummond et al. (2002) show that this works for real data.

Now, if we knew the ancestral sequence B, the final sequence B' and the mutation rate μ we could estimate the time t between the initial sequence and the final sequence. The likelihood for t would be

$$\Pr\{B'|B, \mu, t\} = \prod_{s=1}^{L} [\exp(\mu Q t)]_{B_s, B'_s}. \tag{7.1}$$

We do not have the ancestral sequence B, but we do know something about B, even before we see B'. The sequence B has itself evolved from a sequence of great antiquity. The proportions of **A**'s, **C**'s, **G**'s and **T**'s in B are determined by the relative rates, $Q_{A,C}$ etc., at which these bases mutate into one another. It follows that each character in B is a draw from the equilibrium of the mutation process. Let $\pi = (\pi_A, \pi_C, \pi_G, \pi_T)$ denote this equilibrium distribution (so, $\pi Q = 0$). Since the mutation process acts independently at each site, the probability distribution for B itself is $\Pr\{B\} = \prod_s \pi_{B_s}$.

Of course, knowing B is a draw from the equilibrium of the mutation process doesn't help us pin down t. The likelihood obtained by summing out the unknown B gives us back the equilibrium distribution for B', independent of t. We need to know more about B and we can learn that by looking at its other descendants.

7.3 The Coalescent-Clock

The continuous time model described in this section, called the Kingman coalescent, is described in two classic papers, Kingman (1982a) and Kingman (1982b). The process is extended to serial times in Rodrigo and Felsenstein (1999).

Consider a population of N_e individuals reproducing asexually. We will assume that the population size is constant in time (this assumption may be replaced by any other assumption which reasonably restricts the set of allowed population size histories, for example, to exponential growth at an unknown rate). Consider a pair of generations, and suppose the ith individual in generation one produces n_i offspring in generation two. We model the evolution of a genealogy in the following way. We suppose each individual in generation two chooses its parent uniformly at random from the individuals in generation one, and independent of the choices made by its peers in generation two. This model, which is equivalent to imposing a multinomial distribution for the vector $(n_1, n_2, \ldots, n_{N_e})$ of family sizes, is called the Wright–Fisher population model.

In order to simulate $K - 1$ generations of an ideal Wright–Fisher population, take a piece of paper and mark a square lattice of dots, N_e dots across by K dots up. See Figure 7.1. Connect each dot by a directed edge pointing to a randomly chosen dot in the row above. The "present" generation is the row at the bottom of the lattice, and the earliest is the row at the top. A directed edge leads away from each dot in the array. A lineage is a directed sequence of edges.

Now pick two individuals from the population in the present without reference to their ancestry (i.e. pick two dots from the bottom row without looking at the edges connected to them). Consider the number, M say, of generations back to their most recent common ancestor. The Wright–Fisher model determines a geometric distribution for M, so that $\Pr\{M = m | N_e\} = p \times (1-p)^{m-1}$ with $p = 1/N_e$ (p is the probability that two individuals have a common parent in the previous generation, so $p(1-p)^{m-1}$ is the probability that the lineages don't coalesce for $m - 1$ generations, and then do coalesce in the mth). Notice that the mean number of generations back to the common ancestor of two individuals in the present is just N_e, the population size.

When N_e is large, the distribution of the time t back to coalescence is well approximated by an exponential. Recall that ρ is the number of years per generation. Let λ be the rate, with units *coalescent events per year*, defined by $p = \lambda\rho$. Now m generations is $t = m\rho$ years, so when $0 < \rho \ll t$ and $N_e \gg 1$, we have $(1 - 1/N_e)^{t/\rho} \simeq \exp(-t/N_e\rho)$ and $t \sim \text{Exp}(\lambda)$. Readers familiar

Wright–Fisher population   Wright–Fisher genealogy   Coalescent genealogy

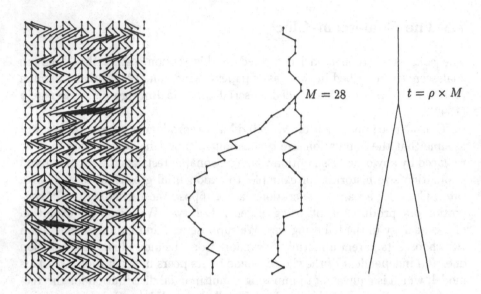

Fig. 7.1. The Wright–Fisher and Kingman coalescent processes, with generation time ρ. (left) $K = 40$ generations of a Wright–Fisher population of $N_e = 20$ individuals, (centre) the ancestral lineages of contemporary individuals 5 and 15 in the population at left coalesce $M = 28$ generations back, (right) the continuous time coalescent tree summarizing the discrete time ancestry in the centre.

with population genetics should note that our λ (which equals $1/(N_e\rho)$) is equivalent to the expression "$1/\Theta$" found elsewhere in the literature.

Now consider what happens if we select not two, but n individuals. Also, we may select individuals from different generations. The ancestral lineages form a tree, with n leaf nodes and $n - 1$ ancestral nodes, corresponding to $n - 1$ coalescent events. At the root of the tree is the node corresponding to the most recent common ancestor of all n individuals at the leaf nodes. As we trace back from the present to the root, the lineages arising from the leaves coalesce one by one until there are just two lineages, which coalesce at the root. The common ancestral tree of n individuals is called their genealogy, and denoted g. It is a tree graph with distinguishable leaf vertices. The tree is drawn upside down, so that the altitude of a vertex is proportional to its age.

The coalescent process we have described determines a probability distribution for g, that is, a probability density for the joint distribution of the tree coalescent times and topology. If we follow any particular pair of lineages back in time, and ignore the rest, the pair we are following behave according to the rule we worked out for $n = 2$; they coalesce at instantaneous rate λ.

With this one rule (and a bit of notation) we can write down the probability density for any particular tree g. As we trace back from the most recent leaf to the root, the number of lineages decreases by one at each coalescent event, and increases by one at each leaf. The number of lineages is a constant in each interval of time between consecutive vertices of the tree.

Imagine simulating the tree from its leaves up to the root. Number the vertices of the tree in order by age from $i = 1$ up to $i = 2n-1$ at the root and assign the vertices ages t_i. Suppose k_i lineages are present in interval $[t_i, t_{i+1})$. The rate R_i for coalescent events in interval i is a constant, $R_i = k_i(k_i-1)\lambda/2$ (the number of distinct pairs multiplied by the rate for each pair). If interval i ends with a coalescence event then its length, $\tau_i = t_{i+1} - t_i$ say, is an exponential variate with mean $1/R_i$. The pair of lineages which coalesce at the top of the interval is chosen from $k_i(k_i - 1)/2$ pairs, so the probability density for that coalescent event was $\lambda \exp(-R_i\tau_i)$. If interval i ends with a leaf then there was no coalescence event in that interval, and that happens with probability $\exp(-R_i\tau_i)$. These events are independent so the probability density $f_G(g|\lambda)$ for the whole tree is a product. Each interval contributes a factor like $\exp(-R_i\tau_i)$ and the $n - 1$ intervals terminated by coalescent events each contribute an extra λ factor. The probability density for the joint distribution of tree coalescent times and topology is then

$$f_G(g|\lambda) = \lambda^{n-1} \prod_{i=1}^{2n-2} \exp\left(\lambda\frac{k_i(k_i - 1)}{2}(t_{i+1} - t_i)\right). \tag{7.2}$$

We return now to the discussion we began in the last two paragraphs of Section 7.2. The mutation model gives us a likelihood for the age of an organism if we know something about the DNA sequences of that organism and its descendants, and the mutation rate. Evidence for the DNA sequence of the ancestral organism is obtained by putting the DNA sequences of its descendants together with prior information about the likely tree structure. The Wright–Fisher population model determines a prior for trees. So, we can expect to be able to date coalescent events if we have DNA sequences for individuals at the leaves of the genealogy, and know the two parameters μ and λ. Can we reconstruct and date a genealogy from the DNA sequences of its leaf individuals if we have the prior and observation models, but the two rate parameters and the genealogy itself are unknown?

7.4 Inference for Rate Parameters

In this section we write down the posterior probability distribution for those unknown parameters of interest which may be estimated from the sequence data of n individual organisms. Exact sequence ages are assumed. We defer treatment of radiocarbon calibration to Section 7.6.

It is convenient at this point to drop the time ordering of vertex labels in g. We will want to make small independent changes to the ages of vertices of g, and we would like them to keep their names as we vary their ages. Let I [Y] denote the set of leaf [ancestral] node labels. Let $t(g) = (t_1, t_2, \ldots, t_{2n-1})$ where t_i is the age of vertex i in genealogy g. Split the vector $t(g)$ into two vectors, $t_I = (t_{I_1}, t_{I_2}, \ldots, t_{I_n})$ and $t_Y = (t_{Y_1}, t_{Y_2}, \ldots, t_{Y_{n-1}})$. Let $R \in Y$ denote the label of the root node. Let $E(g)$ denote the edge set of g, with the convention $\langle i, j \rangle \in E(g) \Rightarrow t_i \geq t_j$. A genealogy is determined by its edge set and vertex times, $g = (E, t)$. Let B be a $(2n-1) \times L$ array of DNA characters. A row of B corresponds to a DNA sequence. Let $B_{i,:}$ denote the ith row of B and $B_{i,s} \in \mathcal{C}$, ($\mathcal{C} = \{\mathbf{A}, \mathbf{C}, \mathbf{G}, \mathbf{T}\}$) denote the character at site s in the DNA sequence for vertex i. Let B_I and B_Y denote the sub-arrays of leaf and ancestral node sequences respectively.

We can think of the coalescent process, which determines the tree-genealogy g, as laying down the railway tracks, along which the mutation process runs. The root sequence is drawn from the equilibrium of the mutation process. The transition probability $\Pr\{B_{j,s} = b | B_{i,s} = a, \mu, t_i - t_j\}$ of Equation 7.1 carries the sequence down from one node to the next down the tree. The probability $\Pr\{B_I, B_Y | g, \mu\}$ for the mutation process acting over tree g to generate ancestral sequences B_Y and leaf data B_I is then

$$\Pr\{B_I, B_Y | g, \mu\} = \prod_{s=1}^{L} \pi_{B_{R,s}} \prod_{\langle i,j \rangle \in E(g)} \left[\exp\left(Q\mu(t_i - t_j) \right) \right]_{B_{i,s}, B_{j,s}} . \qquad (7.3)$$

Our data D are the n dated sequences $D = \{t_i, B_{i,:}\}, i \in I$. The tree topology, E, the $n-1$ undated sequences $\{t_j, B_{j,:}\}, j \in Y$ and the mutation and coalescent rate parameters μ and λ are unknown. Let $x = (\mu, \lambda, E, t_Y, B_Y)$ denote the set of unknowns in this problem. Our inference is based on the posterior probability density $f_{X|D}(x | B_I, t_I)$. Suppose a prior density $p(\mu, \lambda)$ is given. We write the posterior as a product of the conditional probabilities determined by the mutation and coalescent processes,

$$f_{X|D}(x | B_I, t_I) \propto \Pr\{B_I, B_Y | g, \mu\} \, f_G(g | \lambda) \, p(\mu, \lambda). \qquad (7.4)$$

Note that we write g where E, t_I and t_Y appear together.

We now discuss the problem of deciding a prior for μ and λ. This is a density on just two variables, both of which are scale parameters. There should be little mystery in the business. The difficulties treated below arise because we choose to illustrate our methods for a diffuse prior, so we take a paragraph to justify this choice. First, we wish to establish sampling methodology,

and the MCMC sampling problems we consider become more difficult as we use a more diffuse prior. Secondly, any careful Bayesian inference must make some model sensitivity analysis, and a straightforward way to do this is to probe the data with more and less informative priors. For that reason the diffuse priors we consider here may be of use in a more informed analysis.

Thirdly and finally, naive application of a diffuse prior leads to an improper posterior for the parameters of interest. It is useful to warn against this error, and to identify readily available scientific knowledge which is sufficient to fill the gap.

In Drummond et al. (2002) we show that a fairly straightforward MCMC scheme is adequate for inferential problems of real interest. Earlier work (Kuhner et al. 1995) treated the estimation of λ and g if μ was known, and of μ and g if λ was known. Drummond et al. (2002) estimate μ, λ and g jointly, a substantially harder problem.

Joint estimation is not possible if the data are exactly contemporaneous, that is, if $t_i = t_j$ for all $i, j \in I$. Shift the zero of time so that $t_i = 0, i \in I$ and fix a real constant $c > 0$. Consider Equation 7.4 and the transformation $x \to cx$ defined by $cx = (\mu/c, \lambda/c, E, ct_Y, B_Y)$. The rates go down as the tree depth goes up. The factors $\Pr\{B_I, B_Y | g, \mu\}$ and $f_G(g|\lambda)d^{n-1}t_Y$ are invariant under this transformation. The only c-dependence left in the posterior is in the prior distribution for μ and λ. In other words, the data tell us nothing about c that we didn't already know. Here is another way to think about the problem. The transformation $x \to cx$ does not scale the leaf node times, just the ancestral node times, so it is not in general simply a change in the units of time. However, for equal-time leaves $x \to cx$ is indistinguishable from a change of units for time since the leaves are at time zero, and c times zero is zero.

This argument does not go through when leaf vertices are not all contemporaneous. In Equation 7.4, c does not cancel in factors involving edges connecting leaf and ancestral nodes. The time offset between leaves makes one time-scale special, and time-scale invariance is lost in Equation 7.4. However, although the qualitative property of identifiability is present whenever leaf times are not exactly contemporaneous, we can expect a great deal of uncertainty in the scale factor c when the leaf spacing is slight. In particular, if we take c very large, the spacing between leaves is small compared with the time-scale for events in the tree, and the identifiability problem present for equal time-leaves reappears.

Warning   The posterior density $f_{X|D}$ of Equation 7.4 is improper for $p(\mu, \lambda)$ proportional to $1/(\mu\lambda)$.

The warning tells us that "naive" non-informative inference is not possible for the joint estimation of μ, λ and g. A proof of this result is given in Appendix A. The basic problem is that the factors $\Pr\{B_I, B_Y | g, \mu\}$ and $f_G(g|\lambda)$ do not go to zero sufficiently fast as $c \to \infty$ to yield a finite integral over g, μ and λ.

What state of knowledge does determine a proper posterior? Of course this is in a certain sense straightforward. Any proper $p(\mu, \lambda)$ will do the job and in most applications a very little elicitation will determine such a prior. However, as we mentioned above, it is useful, for the purpose of sensitivity analysis, and for challenging MCMC algorithms, to consider very diffuse priors. It is appealing to biologists to allow states at $\mu \to 0$ and $\lambda \to 0$, at least in exploratory analysis. However, a conservative bound t_R^* on the maximum age of the root is readily approved.

> **Encouragement**   Suppose the data B_I contain at least two non-identical sequences. Let positive constants μ^*, λ^* and t_R^* be given. The posterior density $f_{X|D}$ of Equation 7.4 is a proper probability density if $p(\mu, \lambda)$ is proportional to $1/(\mu\lambda)$ and the conditions $\mu < \mu^*$, $\lambda < \lambda^*$ and $t_R < t_R^*$ apply.

This result is established in Appendix B. Why is it worth stating? We saw that integration of $f_{X|D}$ along any ray $\{(\mu/c, \lambda/c, ct_Y); c > 0\}$ is undefined. Divergences arise in two additional limits, $\mu \to \infty$ and $\lambda \to \infty$. These wildly unphysical states cause problems. Firstly

$$\lim_{\mu \to \infty} \Pr\{B_I, B_Y | g, \mu\} = \prod_{s=1}^{L} \prod_{i=1}^{2n-1} \pi_{B_{i,s}}.$$

At high mutation rates there is essentially an instantaneous mutational equilibrium, and the likelihood goes to a non-zero constant corresponding to the equilibrium base frequencies. Second, the coalescent density $f_G(g|\lambda)$ concentrates on very short trees as $\lambda \to \infty$, giving rise to a non-integrable divergence. In terms of the original Wright–Fisher model, $\lambda = 1/(N_e\rho)$, so this divergence arises in the zero population limit. Before we do anything else, we must eliminate these states, and we do that with $\mu < \mu^*$, $\lambda < \lambda^*$. We are now ready to deal with the divergence along rays $\{(\mu/c, \lambda/c, ct_Y); c > 0\}$. These rays are truncated by the root age bound $t_R < t_R^*$. Since the space of states is now closed and bounded, any bounded prior gives a proper posterior, so our encouragement refers only to the scale invariant prior, $1/(\mu\lambda)$.

Notice that a maximum-likelihood estimate of λ and μ, from the integrated likelihood surface $\Pr\{B_I | \mu, \lambda\}$, can be made using draws from $f_{X|D}$. An estimate of the marginal posterior surface for μ and λ becomes an estimate of the likelihood surface $\Pr\{B_I | \mu, \lambda\}$ (up to an overall normalization) by simply dividing out by $p(\mu, \lambda)$.

7.5 Pruning

Before we continue, we note that we have the option, in this inference, to sum out the (typically) uninteresting unknown ancestral sequences B_Y and work with state $x = (\mu, \lambda, E, t_Y)$ and likelihood $\Pr\{B_I | g, \mu\}$. In this section we write down an efficient algorithm, invented by Felsenstein (1981), for this task.

We may compute the sum $\Pr\{B_I | g, \mu\} = \sum_{B_Y} \Pr\{B_I, B_Y | g, \mu\}$ over all $B_Y \in \mathcal{C}^{(n-1) \times L}$ numerically, without recourse to Monte Carlo, using the pruning algorithm of Felsenstein (1981). Vertex j is a child vertex of vertex i in

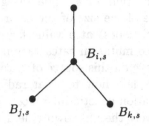

Fig. 7.2. A subtree with characters $B_{i,s}$, $B_{j,s}$, $B_{k,s}$ at site s of the sequences at vertices i, j, and k respectively.

genealogy g if there exists an edge $\langle i,j \rangle \in E(g)$. Suppose vertex i has child vertices j and k (see Figure 7.2). For $a \in C$, the likelihood $\Pr\{B_I|g, \mu, B_{i,s} = a\}$ can be written in terms of the corresponding likelihoods evaluated at j and k,

$$\Pr\{B_I|g,\mu, B_{i,s} = a\} = \sum_{b \in C} [\exp(\mu Q(t_i - t_j))]_{a,b} \Pr\{B_I|g,\mu, B_{j,s} = b\}$$

$$\times \sum_{b' \in C} [\exp(\mu Q(t_i - t_k))]_{a,b'} \Pr\{B_I|g,\mu, B_{k,s} = b'\}.$$

The likelihood at the root $\Pr\{B_I|g,\mu, B_{R,s} = a\}$ is defined by a recursion of the above expression down the tree to the leaves. Let $\mathbb{I}_{\mathcal{E}}$ denote the indicator function for the event $\mathcal{E}$. If j is a leaf then $j \in I$ and $\Pr\{B_I|g,\mu, B_{j,s} = b\} = \mathbb{I}_{B_{j,s}=b}$. There are two sums over four elements at each level of the recursion. The integrated likelihood is given in terms of the equilibrium frequencies, π, defined in Section 7.2, by

$$\Pr\{B_I|g,\mu\} = \prod_{s=1}^{L} \sum_{b \in C} \pi_b \Pr\{B_I|g,\mu, B_{R,s} = b\}.$$

7.6 Uncertainty in Fossil Sequence Dates

There is no discussion, in the literature to date, of the likely impact of radiocarbon calibration errors on genetic inference. However, in paired studies which we omit from the present work, in which we alternately treat and ignore the uncertainty due to radiocarbon calibration, we find that the effect is not important (so, for example, Lambert et al. 2002, are correct to ignore the issue). By far the greatest part of the uncertainty in reported ages comes from the uncertainty in the rate parameters and genealogy (and, no doubt, model mis-specification error). There are around $(\mu L)^{-1}$ years (i.e. about 2000 years in Section 7.8) between mutations on a single lineage. This sets a lower bound on the order of magnitude of the size of the error bars for all age estimates (both leaves and ancestral nodes) at a value far above typical radiocarbon uncertainty. However, since mutation rates vary from species to species, researchers should keep an eye on this source of uncertainty.

In the following we explain how to treat radiocarbon calibration as an explicit part of the population-genetic inference. For each leaf $i \in I$, let T_i, y_i and σ_i denote respectively the unknown true age, the conventional radiocarbon age, and measurement error associated with DNA sequence $B_{i,:}$. Let $d(\tau)$ denote the radiocarbon calibration curve with age dependent error $\sigma(\tau)$ as published in Stuiver et al. (1998). We fit the standard radiocarbon observation model (described, for example, in Buck et al. 1991) to the data y_I, that is

$$y_i \sim d(T_i) + \epsilon(T_i) + \epsilon_i,$$

where $\epsilon(T_i) \sim N(0, \sigma(T_i)^2)$ and $\epsilon_i \sim N(0, \sigma_i)$ are unknown additive Gaussian noise variates. In the absence of sequence data, the posterior density $f_{T_I}(t_I|y_I) \propto f_{Y_I|T_I}(y_I|t_I) f_{T_I}(t_I)$, for the calibrated ages t_I is given in terms of a radiocarbon likelihood $f_{Y_I|T_I}$ and a prior f_{T_I}.

We elicit a prior density f_{T_I} as follows. Suppose the effective population $N_e(\tau)$ is a function of age τ. Suppose that age termini A and P are available, so that for $i \in I$, $A \le t_i \le P$ is prior knowledge, but that otherwise each data-sequence $B_{i,:}$ might equally well belong to any individual in the population history from P to A. The state of knowledge described above is, therefore, represented by a prior density

$$f_{T_I}(t_I) = \prod_{i \in I} \frac{N_e(t_i)}{\int_A^P N_e(\tau) d\tau}$$

defined for $t_I \in [A, P]^n$ (recall, n leaf labels in I). Where λ is estimated as a function of time, it will be necessary to model the action of taphonomy and specimen selection on recovered fossil DNA. In what follows we assume N_e is a constant and ignore selection due to taphonomy. In this setting the above considerations lead to $f_{T_I}(t_I) \propto 1$, the constant prior. This form is used throughout the radiocarbon literature, from Buck et al. (1991) onwards. We choose it, in the example which follows, not because we are reaching for some default, non-informative prior, but because it is computed from a simple explicit model (of the kind described in Nicholls and Jones 2001) of the processes which realize the parameters in question. See Drummond et al. (2002) for a discussion of the age-dependent case.

We must modify Equation 7.4 to take into account the uncertainty arising from the calibration. Our data D are the n dated sequences $D = \{y_i, B_{i,:}\}, i \in I$. Let $x = (\mu, \lambda, E, t_I, t_Y, B_Y)$ denote the set of unknowns in this problem. The leaf times t_I have joined the set of unknowns. The revised posterior is

$$f_{X|D}(x|B_I, y_I) \propto \Pr\{B_I, B_Y|g, t_I, \mu\} \, f_G(g|t_I, \lambda)$$
$$\times \, f_{Y_I|T_I}(y_I|t_I) \, f_{T_I}(t_I) \, p(\mu, \lambda). \qquad (7.5)$$

Notice that undated leaves introduce the possibility of an improper posterior, since the likelihood for an undated leaf, attached at an age, τ say, greater than the root age t_R, does not go to zero as $\tau \to \infty$. Some upper bound on the leaf (or root) age must be provided as prior knowledge. If the data are sufficiently informative it is possible to set this upper bound to an extremely conservative value. The mass of probability in the upper tail of the age distribution is then negligible. This is the case in the example of Section 7.8.

7.7 MCMC

We have implemented an MCMC algorithm generating $X \sim f_{X|D}$. In fact we made three more or less independent implementations of the entire MCMC scheme. One, in MatLab, does not represent sequences B_Y on ancestral vertices, using the above pruning scheme to eliminate those variables. This first implementation samples the marginal posterior distribution for μ, λ and g obtained by summing B_Y out of $f_{X|D}$. A second implementation, also in MatLab, does represent the ancestral sequences in the state, placing them on an equal footing with μ, λ and g in the Monte Carlo. A third Java implementation uses pruning. The multiple implementations were used for checking and debugging, and to investigate the relative efficiency of pruning. Pruning proved to be particularly helpful at high mutation rates, where there is real uncertainty in the ancestral sequences.

We describe in Appendix C a collection of MCMC updates for the case where ancestral sequences are an explicit part of the MCMC simulation. We give details for those updates which are in our opinion difficult to compute, or interesting in other respects. Updates for the two implemented MCMC samplers which use pruning are described in Drummond et al. (2002). Updates of the kind discussed below, in particular, updates which treat sequences at ancestral nodes explicitly, may be found elsewhere, for example, Wilson and Balding (1998) and Wilson et al. (2003). An even more "explicit" treatment may be found in Beaumont (1999), where individual mutation events are represented. Wilson and Balding (1998) and Beaumont (1999) treat microsatellite data, as opposed to DNA sequence data.

For readers who wish to apply the methods described in this chapter to data sets of their own, the MEPI software makes the business as straightforward as one could reasonably hope (http://www.cebl.auckland.ac.nz/mepi/).

7.8 Example

By way of example, we consider a synthetic problem set up to resemble the problem treated in Lambert et al. (2002). We allow for the uncertainty arising from the simultaneous estimation of mutation and coalescent rates and genealogy (μ, λ and g), and from the calibration of radiocarbon-dated mtDNA sequences. Apart from Drummond et al. (2002), Lambert et al. (2002) is the only published analysis to take into account uncertainty arising from the simultaneous estimation of μ, λ and g. We illustrate "genetic dating" of leaf sequences and common ancestors.

Lambert et al. (2002) treat fossil sequences from penguins. They sequenced the mitochondrial HVRI region using material from 96 ancient bone samples, up to around 6500 years in age, and 380 blood samples from modern birds at 13 Antarctic locations. Their analysis is based on 352 aligned sites in the 96 fossil sequences and an unpublished subset of the modern sequences. We simulate $n = 22$ sequences of length $L = 400$ with no gaps. We leave two sequences (from the middle and end of the genealogy) completely undated, in order to illustrate genetic dating. The dated sequences in the data allow us to say something about the unknown ages of the undated sequences. Lambert et al. (2002) do not publish an estimate of the effective population size. We suppose $N_e = 1000$, an order of magnitude for populations of this sort. They assume a generation time of $\rho = 5.5$ years and estimate a mutation rate of around 10^{-6} mutations per site per year.

Let Λ, M, G and T_I denote the synthetic true coalescent and mutation rates, synthetic true genealogy and leaf times. In line with Lambert et al. (2002), we choose $\Lambda = 1/5500$ and $M = 10^{-6}$. We distribute the 22 true leaf times T_I uniformly between the present and $11000\,\text{BP}$, and, for $i \in I$, simulate synthetic radiocarbon data $y_i \sim f_{Y|T}(\cdot|T_i)$. Synthetic sequence data are drawn by simulating a true tree $G \sim f_G(\cdot|\Lambda, T_I)$ (see Figure 7.3 (top plot)), drawing synthetic root characters $B_{R,s} \sim \pi$ for each $s = 1, 2, \ldots, L$, and then simulating leaf sequences $B_I \sim \Pr\{\cdot|G, M, B_{R,:}\}$ by simulating the mutation process $\Pr\{B_{j,s} = b|B_{i,s} = a, M, t_i - t_j\}$ down each edge $\langle i, j \rangle \in E(G)$ from the root to the leaves. In order to make the inference proper, we impose upper limits, $\mu^* = 1$ mutation per site per year, $\theta^* = 1/5.5$ (so $N_e \geq 1$) and $t_R^* = 40000$. Lower limits are all zero. The first two bounds are almost completely uninformative. In fact the Monte Carlo did not visit any of the bounds. The posterior probability of states in the vicinity of the bounds is negligible.

We carry out sample-based Bayesian inference, simulating $X_k \sim f_{X|D}, k = 0, 1, \ldots, K$ with $K = 5 \times 10^6$ using the MCMC scheme of Appendix C. The MCMC is started with a tree drawn from f_G, the coalescent prior. A tree g, sampled from the posterior (simply the last tree in the run) is shown in Figure 7.3 (bottom plot). The MCMC output for the slowest mixing statistic (that is, the state function $h(x)$ with the greatest integrated autocorrelation time) can be seen in Figure 7.4 along with its autocorrelation function, and its large-lag asymptotic variance ($\pm 2\sigma$). The run contains about 400 effective independent samples. For details of these convergence diagnostics, see Geyer (1992).

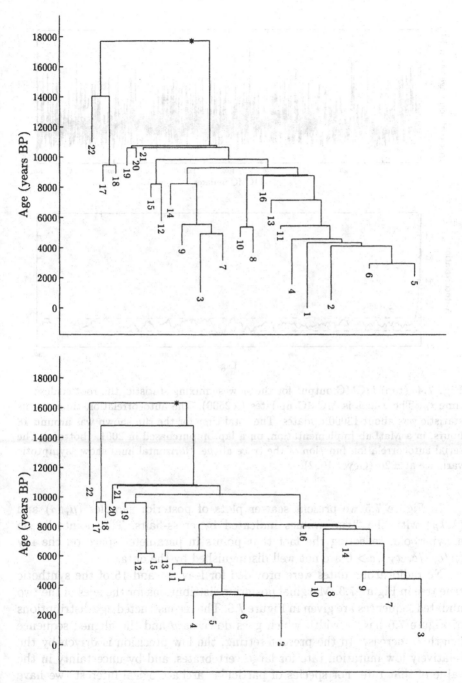

Fig. 7.3. (top) Synthetic data, the true genealogy. Leaf labels correspond to distinct fossil sequences. (bottom) A genealogy sampled from the posterior distribution.

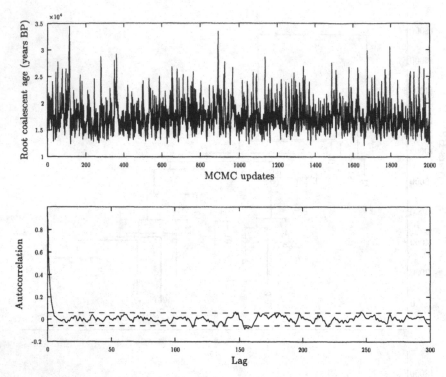

Fig. 7.4. (top) MCMC output for the slowest mixing statistic, the root coalescent time t_R. The x-axis is MCMC updates ($\times 2500$). The autocorrelation time of this statistic was about 13000 updates. The total time for the run shown was around 18 hours, in a MatLab implementation, on a laptop purchased in 2001. (bottom) The serial autocorrelation function of the trace above. Horizontal lines show asymptotic variance at $\pm 2\sigma$ (Geyer 1992).

In Figure 7.5 we present scatter plots of posterior samples (μ, t_R) and (λ, t_R) with the "true" values indicated by cross-hairs. The points lie on a hyperbola, reflecting the fact that points in parameter space on the ray $(\mu/c, \lambda/c, ct_Y), c > 0$ are not well distinguished by the data.

No radiocarbon dates were provided for leaves 1 and 15 of the synthetic true tree in Figure 7.3. Marginal posterior distributions for the ages of the two undated sequences are given in Figure 7.6. The reconstructed age distributions of Figure 7.6 have a width which goes down as μ and the aligned sequence length L increase. In the present setting, the low precision is driven by the relatively low mutation rate for large vertebrates, and by uncertainty in the value of that rate. For species of particular archaeological interest (we have in mind the Polynesian rat) some of this uncertainty could be removed using independent measurements of mutation rates.

The model we are fitting makes a number of assumptions which are unlikely to hold for this population. The real population size is not constant.

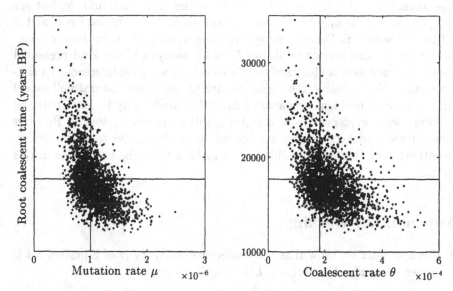

Fig. 7.5. Scatter plots of posterior samples of (μ, t_R) (left) and (λ, t_R) (right). Cross-hairs indicate true parameter values. Note: $\lambda = 1/(N_e \rho)$, see Section 7.3 for details.

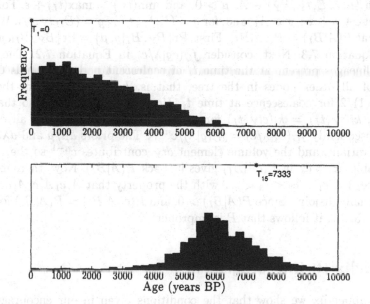

Fig. 7.6. Ancient or modern? Marginal posterior distributions for the ages of undated mtDNA sequences: (top) leaf 1 of Figure 7.3, (bottom) leaf 15. Synthetic true ages are indicated with an asterix.

The animals are distributed in breeding colonies which intermingle, but are not panmictic. The mutation process is correlated along the sequence, and is subject to selection. Because we ignore these properties, immediate chronometric conclusions cannot be drawn from an analysis of the kind presented above. The aim here is illustrate sample-based genealogical inference in a simple setting. Nevertheless, the model we are fitting is the natural null model for this kind of problem. Departures from this model may be expected, and in future work evidence for such features will no doubt be sought. However those future model comparison studies will be made relative to this model, or something very similar, and will need to make a fit of the kind made in this section.

Appendix A: Warning

In this appendix we show that the posterior density $f_{X|D}$ of Equation 7.4 is improper when $p(\mu, \lambda) \propto 1/(\mu\lambda)$. Let

$$P(A|B_I) = \int_{A \cap \Omega} f_{X|D}(x|B_I)dx.$$

Let $\epsilon > 0$ and a compact subset A of Ω be given, satisfying $P(A|B_I) > 0$ and, for each $(\mu, \lambda, E, t_Y, B_Y) \in A$, $\mu > 0$, and $\min(t_Y) > \max(t_I) + \epsilon$. For each $c > 1$ let $cA = \{cx : x \in A\}$, and, for $g = (E, t_Y, t_I)$, $cg = (E, ct_Y, t_I)$. We now show that $P(A|B_I) = P(cA|B_I)$. First, $\Pr\{B_Y, B_I|g, \mu\} = \Pr\{B_Y, B_I|cg, \mu/c\}$ from Equation 7.3. Next, consider $f_G(cg|\lambda/c)$ in Equation 7.2. The number of lineages present at the time t_i of coalescent node i depends on the times of all other nodes in the tree, that is, $k_i = k_i(t_Y, t_I)$ in the rate $k_i(k_i - 1)/2$ for coalescence at time t_i. However, A is defined so that, for $x \in A$, $k_i(t_Y, t_I) = k_i(ct_Y, t_I)$ for all $i = 1, 2, \ldots, 2n - 1$ and all $c > 1$, and consequently $f_G(cg|\lambda/c) = f_G(g|\lambda)/c^{n-1}$. The priors $d\mu/\mu$ and $d\lambda/\lambda$ are scale invariant, and the volume element dt_Y contributes c^{n-1} so the change of variables $x' = cx$ in $P(cA|B_I)$ gives us back $P(A|B_I)$. Now, there exists a sequence, $1 < c_1 < c_2 < c_3 < \ldots$ with the property that $A, c_1 A, c_2 A, c_3 A, \ldots$ are mutually disjoint. Since $P(A|B_I) > 0$, and $P(c_n A|B_I) = P(A|B_I)$ for each $n = 1, 2, 3, \ldots$, it follows that P is improper.

Appendix B: Encouragement

In this appendix we show that the conditions given in our encouragement in Section 7.4 determine a proper posterior. First we bound $\Pr\{B_Y, B_I|g, \mu\}$ away from one, for $t_R \leq t_R^*$. Since the B_I are not identical, there is at least one mutation over g. The probability to get the sequences B_I and B_Y is less than the probability that there is at least one mutation (since B_I implies a mutation), which is one minus the probability for no mutations on g,

$$\Pr\{B_Y, B_I | g, \mu\} \leq 1 - \sum_{b \in C} \pi_b \exp\left(\mu Q_{b,b} | g|\right)$$

$$\leq 1 - \sum_{b \in C} \pi_b \min_{b \in C} \left(\exp\left(\mu Q_{b,b}(2n - 2)t_R^*\right)\right)$$

where $|g| = \sum_{\langle i,j \rangle \in E(g)} |t_i - t_j|$ is the total edge length. Note that $Q_{b,b}$ is negative and $|g| \leq (2n - 2)t_R^*$, as the greatest tree length is less than the number of edges times the maximum edge length. Let Γ denote the set of all genealogies g allowed for leaf times t_I and given $t_R \leq t_R^*$. Integration dg involves integration $d^{n-1}t_Y$ and summation over all distinct tree topologies E. The normalizing constant $Z = \int f_{X|D} dx$ is

$$Z = \int_0^{\mu^*} \int_\Gamma \int_0^{\lambda^*} \left[\sum_{B_Y \in \mathbb{B}} \Pr\{B_Y, B_I | g, \mu\} \right] f_G(g|\lambda) p(\lambda, \mu) d\lambda dg d\mu$$

$$\leq 4^{(n-1)L} \int_0^{\mu^*} \frac{1}{\mu} \left(1 - \min_{b \in C} \left(\exp\left(\mu Q_{b,b}(2n-2)t_R^*\right)\right) \right) d\mu$$

$$\times \int_\Gamma \int_0^{\lambda^*} \lambda^{n-2} \exp\left(-\frac{\lambda}{2} \sum_{i=1}^{2n-2} k_i(k_i - 1)(t_{v(i)+1} - t_{v(i)}) \right) d\lambda dg.$$

The integral over μ is finite. The integration $d\lambda dg$ is a sum, over tree topologies which are finite in number. Each term of that sum is given by the integral $d\lambda dt_Y$ of a bounded function over a bounded domain. It follows that the posterior is proper.

Appendix C: Markov chain Monte Carlo

In this appendix we give an MCMC scheme for states with ancestral sequences. For our example, we suppose $p(\lambda, \mu) = (\lambda \mu)^{-1}$. Let Ω denote the space of states $x = (\mu, \lambda, E, t_Y, B_Y)$. We specify a Markov chain X_k, $k = 0, 1, 2, \ldots$, with states, $X_k \in \Omega$, and equilibrium $f_{X|D}$. Metropolis et al. (1953), Hastings (1970) and Green (1995) define a class of Monte Carlo update algorithms which determine a transition matrix stationary with respect to a given target distribution.

Suppose $X_k = x$. A value for X_{k+1} is computed using the Metropolis–Hastings algorithm. First, a candidate state x' is generated by randomly perturbing x in some way. An operation of type m is chosen at random from a list $m = 1, 2, \ldots, M$ of operation types. The state x' is generated. This is implemented by drawing uniform random variates $u = (u^{(1)}, u^{(2)}, \ldots)$ according to according to a density $q_m(u)$, and computing some function $x' = x_m(x, u)$. For example, to do a random walk update to μ with constant window size $z > 0$, draw $u \sim U(0, 1)$ and set $\mu' = \mu + z(2u - 1)$. Consider now the reverse operation. Suppose the draw $u' \sim q_m$ maps x' back to x, so that $x = x_m(x', u')$.

In the random walk example, $u' = 1 - u$. Secondly, we accept the candidate, and set $X_{k+1} = x'$ with probability

$$\alpha_m(x'|x) = 1 \wedge \frac{f_{X|D}(x'|B_I, t_I)q_m(u')}{f_{X|D}(x|B_I, t_I)q_m(u)}\left|\frac{\partial(x', u')}{\partial(x, u)}\right|$$

(where $a \wedge b$ equals a if $a < b$ and otherwise b), for update type m. If the candidate is not accepted, we set $X_{k+1} = x$, so the state of the chain is unchanged.

We may choose the proposal scheme $q_m, x_m, m = 1, 2, \ldots, M$ as we please, subject to conditions outlined, for example, in Tierney (1994). The role of the Jacobean factor is clarified in Green (1995) in a general setting. As an example, in the random walk update above, the relevant block of the Jacobian matrix is $\begin{bmatrix} 1 & 2 \\ 0 & -1 \end{bmatrix}$ so the absolute value of the determinant is equal to one and the acceptance probability for the random walk update to μ is $1 \wedge f_{X|D}(x'|B_I, t_I)/f_{X|D}(x|B_I, t_I)$. In our MCMC we need to use an update in which this Jacobian factor is not equal to one. We saw, in Section 7.4, that the posterior density can be expected to possess a ridge, which we may move along using the operator $x' = cx$. The update then is as follows. Suppose $X_k = x$. Choose $c \sim U(1/2, 2)$. Set $x' = (\mu/c, \lambda/c, E, ct_Y, B_Y)$. This may result in $x' \notin \Omega$ (for example, scaled ancestral node ages ct_Y may violate the parent–child age order relation for edges in E). If this is the case x' will be rejected at the next step. If $x' \in \Omega$, the candidate is admissible, and the acceptance probability is $1 \wedge c^{n-5} f_{X|D}(x'|B_I, t_I)/f_{X|D}(x|B_I, t_I)$. Let us see how the factor c^{n-5} arises. Since $t'_Y = ct_Y, \mu' = \mu/c, \lambda' = \lambda/c$ and $c' = 1/c$ (so that $x = c'x'$), $\partial(x', c')/\partial(x, c)$ has diagonal $(c, \ldots [n - 3 \; c\text{'s}] \ldots c, c^{-1}, c^{-1}, -c^{-2})$. The off-diagonal elements are zero, except the last column which contains non-zero elements. The determinant of this matrix is $-c^{n-5}$.

Some of the parameters of the problem may feasibly be Gibbs-sampled (Suomela 1976). For parameter $p \in x$, let $x_{-p} = x \setminus \{p\}$ denote x with p omitted. The conditional density of $\lambda|x_{-\lambda}, D$ in $f_{X|D}$ is

$$\lambda|x_{-\lambda}, D \sim \lambda^{n-2} \exp(-\beta\lambda)\mathbb{I}_{\lambda \leq \lambda^*}.$$

where

$$\beta = \frac{1}{2}\sum_{i=1}^{2n-2} k_i(k_i - 1)(t_{v(i)+1} - t_{v(i)}).$$

Here $v(i)$ is a mapping from the unordered node labels of Section 7.4 back to the age-ordered node labels of Section 7.3. This is just a Gamma$(n - 1, \beta)$ density on $\lambda \leq \lambda^*$.

As discussed above, we have the option to sum out the ancestral sequences B_Y from the posterior distribution, using the pruning algorithm. If that is done, B_Y does not arise in x, the Monte Carlo state. If we choose not to prune, so B_Y is part of x, then we need some MCMC update for the conditional distribution of $B_Y|B_I, g, \mu$ determined by $f_{X|D}$. This conditional distribution is in fact a Markov random field (MRF), in which each of the $n-1$, L-component variables $B_i \in \mathcal{C}^L, i \in Y$ is conditionally independent of the rest, given the sequences at its neighbours i_1, i_2 and i_3 on the tree. The neighbours of vertex i in tree g are those vertices to which it is connected by an edge in $E(g)$. In the update below, i_2 and i_3 are i's child vertices. The root vertex of g is the child of a vertex of infinite age. The MRF may be simulated by the following Gibbs update. Select $i \in Y$ uniformly at random. For each $s = 1, 2, \ldots, L$ and $b \in \mathcal{C}$, calculate the 4-components

$$\mathcal{B}_b = [\exp{(Q\mu(t_{i_1} - t_i))}]_{B_{i_1,s},b} \, [\exp{(Q\mu(t_i - t_{i_2}))}]_{b,B_{i_2,s}}$$
$$\times \, [\exp{(Q\mu(t_i - t_{i_3}))}]_{b,B_{i_3,s}}$$

of the vector $\mathcal{B}(B_{-i,s}, g, \mu) = (\mathcal{B}_\mathbf{A}, \mathcal{B}_\mathbf{C}, \mathcal{B}_\mathbf{G}, \mathcal{B}_\mathbf{T})$. Draw $B_{i,s} = b$ with probability $\mathcal{B}_b/Z_\mathcal{B}$ where $Z_\mathcal{B}(B_{-i,s}, g, \mu) = \sum_{a \in \mathcal{C}} \mathcal{B}_a$. The acceptance probability is equal to one.

It is necessary to have some topology-changing update, so that tree-space is explored. We make some small random modification of the tree topology, so that the new state x' is equal to x up to $E \to E'$. For example, we can choose two edges $\langle i, i' \rangle$ and $\langle j, j' \rangle$ in E and reconnect them as $\langle i, j' \rangle$ and $\langle j, i' \rangle$. If the resulting tree is not admissible the candidate state will be rejected. The probability to generate E' from E in this way is just the probability to chose the two edges by which they differ, so the acceptance probability is $1 \wedge f_{X|D}(x'|B_I, t_I)/f_{X|D}(x|B_I, t_I)$. Where simulation is made with ancestral sequences an explicit part of the Monte Carlo state, we improve the candidate's chances if we draw new sequences at vertices i and j, using the above Gibbs proposal in the new tree. This is not exactly a Gibbs update, since the conditional distributions $\Pr\{B_i|B_{-i}, g', \mu\}$ and $\Pr\{B_i|B_{-i}, g, \mu\}$ involved in the forward and reverse proposals are normalized on different trees. The acceptance probability $1 \wedge \prod_{k=i,j} \prod_{s=1}^L Z_\mathcal{B}(B_{-k,s}, g', \mu)/Z_\mathcal{B}(B_{-k,s}, g, \mu)$ does not quite collapse down to one.

When we account for the uncertainty in the ages of the sequence data, as in Sections 7.6 and 7.8, we need an update varying leaf times t_I, which are otherwise fixed. We use a suite of updates, suggested by our experience in Nicholls and Jones (2001) with MCMC for the posterior distribution of radiocarbon calibration. We omit the Hastings ratios from this chapter as they are simply the constant prior density Hastings ratios of Nicholls and Jones (2001) weighted by the likelihood ratio

$$\frac{\Pr\{B_I, B_Y|g, t_I', \mu\} \, f_G(g|t_I', \lambda)}{\Pr\{B_I, B_Y|g, t_I, \mu\} \, f_G(g|t_I, \lambda)}.$$

We include a number of other move types, including other topology-changing tree operations. We have a random walk move for node age, acting on a single randomly chosen ancestral node. This move generates its candidate by selecting a new time for the node at random between the time of its parent, and oldest child. We have experimented with a wide range of other moves. However, whilst it is easy to think up computationally demanding updates which improve the convergence and mixing rates per Markov chain update, it is harder to find updates that improve the convergence and mixing rates per CPU second. Certain move types which may be of value have not been considered. In particular, updates of the kind described in Mau et al. (1999) which are natural in the cophenetic matrix tree representation, were not considered, though they seem promising.

References

Ammerman, A. J. and Cavalli-Sforza, L. L. (1984). *The Neolithic transition and the genetics of populations in Europe.* Princeton University Press, Princeton.

Barnes, I., Matheus, P., Shapiro, B., Jensen, D. and Cooper, A. (2002). Dynamics of Pleistocene population extinctions in Beringian brown bears. *Science*, **195**, 2267–2270.

Beaumont, M. (1999). Detecting population expansion and decline using microsatellites. *Genetics*, **153**, 2013–2029.

Buck, C. E., Kenworthy, J. B., Litton, C. D. and Smith, A. F. M. (1991). Combining archaeological and radiocarbon information: a Bayesian approach to calibration. *Antiquity*, **65**, 808–821.

Drummond, A. J., Nicholls, G. K., Rodrigo, A. G. and Solomon, W. (2002). Estimating mutation parameters, population history and genealogy simultaneously from temporally spaced sequence data. *Genetics*, **161**, 1307–1320.

Felsenstein, J. (1981). Evolutionary trees from DNA sequences: a maximum likelihood approach. *Journal of Molecular Evolution*, **17**, 368–376.

Geyer, C. J. (1992). Practical Markov chain Monte Carlo (with discussion). *Statistical Science*, **7**, 473–511.

Green, P. J. (1995). Reversible jump Markov chain Monte Carlo computation and Bayesian model determination. *Biometrika*, **82**, 711–732.

Hastings, W. (1970). Monte Carlo sampling methods using Markov chains and their applications. *Biometrika*, **57**, 97–109.

Kingman, J. F. C. (1982a). The coalescent. *Stochastic Processes and their Applications*, **13**, 235–248.

Kingman, J. F. C. (1982b). On the genealogy of large populations. *Journal of Applied Probability*, **19A**, 27–43.

Kuhner, M. K., Yamato, J. and Felsenstein, J. (1995). Estimating effective population size and mutation rate from sequence data using Metropolis-Hastings sampling. *Genetics*, **140**, 1421–1430.

Lambert, D. M., Ritchie, P. A., Millar, C. D., Holland, B., Drummond, A. J. and Baroni, C. (2002). Rates of evolution in ancient DNA from Adelie penguins. *Science*, **195**, 2270–2273.

Matisoo-Smith, E., Roberts, R., Irwin, G., Allen, J., Penny, D. and Lambert, D. (1998). Patterns of prehistoric human mobility in Polynesia revealed by mitochondrial DNA from the Pacific rat. *Proceedings of the National Academy of Sciences of the United States of America*, **95(25)**, 15145–15150.

Mau, B., Newton, M. A. and Larget, B. (1999). Bayesian phylogenetic inference via Markov chain Monte Carlo methods. *Biometrics*, **55**, 1–12.

Metropolis, N., Rosenbluth, A. W., Rosenbluth, M. N., Teller, A. H. and Teller, E. (1953). Equation of state calculations by fast computing machines. *Journal of Chemical Physics*, **21**, 1087–1092.

Nicholls, G. and Jones, M. (2001). Radiocarbon dating with temporal order constraints. *Applied Statistics*, **50**, 503–521.

Penny, D., Watson, E. and Steel, M. (1993). Trees from languages and genes are very similar. *Systematic Biology*, **42**, 382–384.

Rambaut, A. (2000). Estimating the rate of molecular evolution: incorporating non-contemporaneous sequences into maximum likelihood phylogenies. *Bioinformatics*, **16**, 395–399.

Rodrigo, A. G. and Felsenstein, J. (1999). Coalescent approaches to HIV population genetics. In K. Crandall (ed.), *Molecular evolution of HIV*, Johns Hopkins University Press, Baltimore, 233–272.

Stuiver, M., Reimer, P. J., Bard, E., Beck, J. W., Burr, G. S., Hughen, K. A., Kromer, B., McCormac, F. G., Plicht, J. V. D. and Spurk, M. (1998). IntCal98 radiocarbon age calibration, 24,000-0 cal BP. *Radiocarbon*, **40**, 1041–1083.

Suchard, M. A., Weiss, R. E. and Sinsheimer, J. S. (2001). Bayesian selection of continuous-time Markov chain evolutionary models. *Molecular Biology and Evolution*, **18**, 1001–1013.

Suomela, P. (1976). *Construction of nearest neighbour systems*. Ph.D. thesis, Department of Mathematics, University of Helsinki.

Tierney, L. (1994). Markov chains for exploring posterior distributions. *Annals of Statistics*, **22**, 1701–1728.

Underhill, P. A., Passarino, G., Lin, A. A., Marzuki, S., Cavalli-Sforza, L. L. and Chambers, G. (2001). Maori origins, Y-chromosome haplotypes and implications for human history in the Pacific. *Human Mutation*, **17**, 271–280.

Wilson, I. J. and Balding, D. J. (1998). Genealogical inference from microsatellite data. *Genetics*, **150**, 499–510.

Wilson, I. J., Weale, M. E. and Balding, D. J. (2003). Inferences from DNA data: population histories, evolutionary processes and forensic match probabilities. *Journal of the Royal Statistical Society, series A, Statistics in Society*, **166**, 1–33.

8

Tephrochronology and its Application to Late Quaternary Environmental Reconstruction, with Special Reference to the North Atlantic Islands

Andrew J. Dugmore, Guðrún Larsen, and Anthony J. Newton

Summary. This chapter offers a review of the current state of the art in tephrochronology. This type of chronology building relies on chemically identifiable tephra deposits created by volcanic eruptions. Identifiable tephra can be spread over large areas and are often found on archaeological sites and in lake sediments, peat deposits and ice cores. Thus, tephras can be used to synchronize deposits at a variety of locations and if reliable dates can be obtained for them, tephrochronology can be used to help build chronologies in a range of disciplines. For many years, tephra have been used to provide spot dates at single sites, but much of the potential for 3D reconstructions and spatial analysis of patterns of change through time have yet to be realized. The chapter discusses the potential for development of suitable formal chronology building tools and highlights the kinds of research problems that need to be tackled.

8.1 Introduction

Tephrochronology is a dating technique based on identifying and correlating horizons of pyroclastic ejecta or *tephra* from volcanic eruptions (Thórarinsson 1944, 1981; Shane 2000). At a fundamental level tephrochronology is indivisible from volcanic history; the timing of tephra-forming events is a key part of the history of a volcanic system, and, vice versa, comprehensive volcanic histories are an essential underpinning for an effective tephrochronology (Larsen et al. 1998). In this chapter we focus on the world-class examples provided by Late Quaternary volcanism in Iceland in order to assess general principles of tephrochronology, and the special attributes of tephra that make it particularly relevant for both constructing chronologies and using them to undertake environmental reconstruction.

Tephra are produced over very short timescales, generally hours to days, and they can be dispersed over very large areas; in the North Atlantic region these range from 10^2 km^2 to more than 10^6 km^2 (Figure 8.1; Larsen et al. 1999; Wastegård et al. 2000a,b; Van den Bogaard and Schminke 2002). Individual tephra layers can be used to define *isochrons*, or time-parallel marker horizons (Thórarinsson 1981; Lowe and Hunt 2001). These isochrons can provide both precise and accurate dating control over very large areas. Iceland is the principal source of tephra in the North Atlantic region and these tephras are predominantly composed of volcanic glass shards (Thórarinsson 1944; Haflidason et al. 2000). This glass can have a distinctive composition that can be used, in combination with reference data from the source areas, to identify both the source volcanoes for the tephra, and the eruptions that produced them (Larsen et al. 1999). Silicic eruptions tend to produce the most widely dispersed tephras and those with the most distinctive individual chemistries (Dugmore et al. 1995a; Pilcher et al. 1996; Boygle 1998; Shane 2000). Some basaltic (i.e. poor in quartz and rich in feldspar) eruptions have also produced extensive tephra horizons, most notably during the last glacial to interglacial transition, but through the Late Quaternary as a whole extensive basaltic tephra deposits are significantly less common than silicic (rich in quartz) deposits (Mangerud et al. 1984, 1986; Zielinski et al. 1995; Davies et al. 2001).

The rapid production, distinctive composition and widespread dispersal of tephra to form stratigraphically precise and spatially extensive isochrons create a number of exceptional opportunities for constructing chronologies.

8.2 Principles of Tephrochronology

The essential methodology of Icelandic tephrochronology was established by the pioneering work of Thórarinsson (e.g. 1944, 1958, 1967). The first stage is to identify primary deposits of tephra in stratigraphic profiles close to the source areas (Figures 8.2 and 8.3). Given sufficient stratigraphic resolution, each profile can be used to determine a minimum number of tephra layers and their relative order. Multiple profiles can then be used to map the extent of each tephra layer, and determine the physical characteristics of component particles (e.g. particle composition, size and shape) and layer thickness. Thickness, or *isopach*, maps can then be used to help determine the source of the tephra (Figure 8.3; Thórarinsson 1944, 1967; but also see Walker 1980). A high spatial density of fully logged profiles is needed to create accurate isopach maps (e.g. Larsen et al. 2001). Although tephras are generally well preserved in the aggrading sediment sequences that typify the andisol and histosol soils of Iceland (Arnalds 1999), many parts of the landscape are not appropriate for determining primary fallout thicknesses. This is because tephra layers can be modified by down-slope transport and deposition, and, even on level areas, tephra layer thicknesses may be modified by post-depositional processes such as cryoturbation and the formation of frost hummocks (Dugmore and Buckland 1991). Maps based on hundreds of selected profiles have now been constructed for many of the historical-age and mid–late Late Quaternary tephras (e.g. Thórarinsson 1954, 1958, 1967; Larsen and Thórarinsson 1977; Sparks et al. 1981; Jóhannesson et al. 1981; Larsen 1984; Saemundsson 1991). Tephras

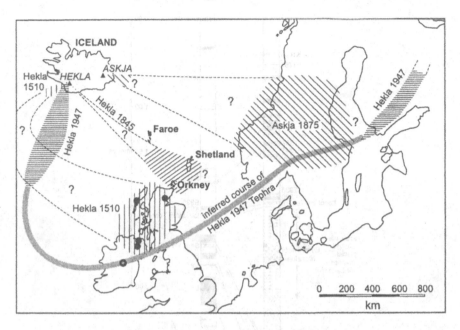

Fig. 8.1. Examples of distal tephra fallout from Icelandic volcanic eruptions of historical age. Fallout from Hekla in AD 1845 is inferred from ships' logs and the analysis of tephra deposits in Orkney (Traill 1845; Connell 1846, and Dugmore, unpublished data). The fallout from Askja in AD 1875 was mapped from contemporary records by Mohn (1877), and fallout data in Finland from the 1947 Hekla eruption are from Salmi (1948). The path of the 1947 volcanic plume was inferred by Thórarinsson (1954), based on data from ships south of Iceland, the eruption column height and the prevailing meteorological conditions. This model is supported by the discovery of the AD 1947 tephra in Ireland close to the putative track of the plume (Hall and Pilcher 2002). Although the plume from the Hekla eruption of AD 1510 initially travelled to the south west (Thórarinsson 1967), the tephra has been recovered from the environmental records of Scotland (Dugmore et al. 1996) and Ireland (Hall and Pilcher 2002). Key sites shown by filled symbols.

have been correlated to source by mapping fall-out from both sub-aerial volcanoes, such as Hekla, and sub-glacial volcanoes, such as Katla, where all the near-vent deposits are either buried by, or have been removed by, the overlying glaciers (e.g. Larsen et al. 2001). Although this mapping approach can clearly identify the source volcanic system for a tephra, and even potential source areas within particular volcanic system, it cannot always be used to identify specific source-vents.

With a sound stratigraphic and spatial framework in place, other characteristics of the tephra can be determined. A particularly distinctive aspect can be the chemical composition of individual glass shards (Larsen 1981). Through geological time, different volcanoes can replicate similar major element compositions. However, through the limited time window of the Late Quaternary,

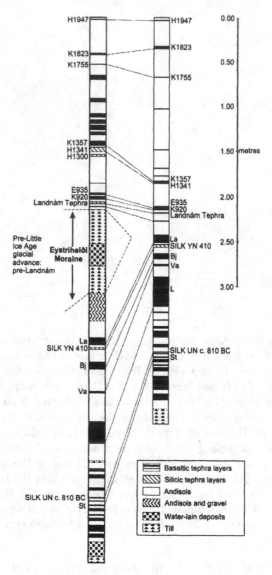

Fig. 8.2. Stratigraphic profiles from the foreland of Sólheimajökull, the principal outlet glacier flowing south-west from Mýrdalsjökull (Figure 8.3). On the left is a composite section through the late prehistoric (6th–9th century AD) Eystriheiði moraines; on the right is a single reference section within 500 m of the moraine limit. In the latter, 32 tephras were deposited in an aeolian sedimentary sequence that developed on a late Holocene till formed about 3100 ^{14}C BP (Dugmore 1989). Rapid accumulation of aeolian material means that tephras formed within a short period are clearly separated; a notable sequence is that of eruptions in AD 1300, 1341 and 1357. Tephras: H: Hekla, K: Katla (basaltic), SILK: Katla (silicic), E: Eldgjá, other letters are defined in Dugmore (1989). Numbers are years AD (e.g. H 1947 is the Hekla tephra from AD 1947), except for the SILK UN tephra of about 810 BC (Dugmore et al. 2000).

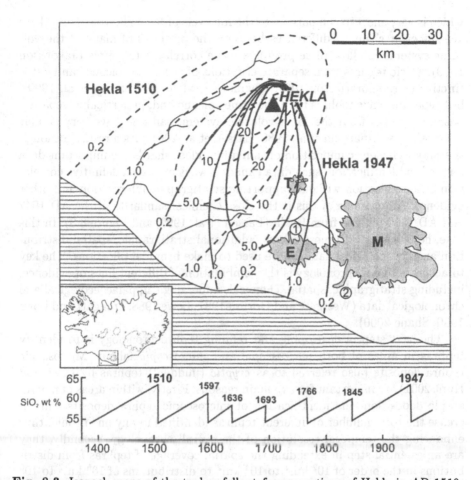

Fig. 8.3. Isopach maps of the tephra fallout from eruptions of Hekla in AD 1510 (dashed lines) and AD 1947 (solid lines); data from Thórarinsson (1954, 1967). The thicknesses are shown in cm, and the location of Hekla is shown by the triangle. Dark shading shows glaciers. A thin deposit (ca. mm) of AD 1510 tephra fell across the small Tindfjallajökull icecap ('T'), but in AD 1947 this icecap was covered with about 5 cm of fallout. In AD 1947 from about 5 cm to less than 1 cm of tephra was also deposited across the larger icecap Eyjafjallajökull ('E'). The largest of the local icecaps, Mýrdalsjökull ('M'), which lies further to the east, was untouched by fallout from the AD 1510 eruption and received only a light dusting of the AD 1947 tephra. The main crater of the Katla volcanic system lies beneath this icecap. Both the AD 1510 and AD 1947 tephras have been recorded in north-west Europe (Figure 8.1). The graph shows the silica composition of the glass fraction of the initial tephra produced by the eruptions of Hekla 1510–1947 (data from Thórarinsson 1967). Although the major element composition of the glass fraction of the tephra is very similar, other data including the five intervening eruptions of Hekla can be used to clearly distinguish the two tephras. '1' indicates the location of the profiles in Figure 8.4, '2' indicates the location of the profiles in Figure 8.2.

which is very short in comparison to the life cycle of Icelandic volcanoes, there are sufficient chemical differences between the products of many of the volcanic systems to allow these products to be correlated to source (Jakobsson 1979). Silicic tephras from separate eruptions of the same volcano can be distinctive (e.g. Thórarinsson 1967; Dugmore et al. 1995a; Larsen et al. 1999), but separate silicic tephras from the same volcano and, in particular, separate basaltic tephras from the same volcanic system, can also have very similar compositions (Einarsson et al. 1980; Larsen et al. 2001). As a result, although the major element composition of individual glass shards are important data for tephra identification, such data cannot always provide a definitive correlation or identification without supporting stratigraphic information and other evidence. An example of this is the close chemical similarity of the AD 1947 and AD 1510 tephras from Hekla (Larsen et al. 1999, and Figure 8.3). In this case, however, as with most others, additional stratigraphic, spatial distribution and chronological data can be used to make firm identifications. The key to accurate tephrochronology is the combination of different lines of evidence, including stratigraphic, spatial, chemical, mineralogical, palaeoecological and chronological data (Westgate and Gorton 1981; Lowe 1988; Froggatt and Lowe 1990; Shane 2000).

The next step in developing and refining tephrochronology is to identify tephras in areas of both *macroscopic* and *microscopic* fallout. Microscopic tephra deposits (also referred to as *cryptic* ('hidden') tephras by Lowe and Hunt 2001) are important for two main reasons. Firstly, within areas of macroscopic deposition, the identification of microscopic tephra deposits will increase the total number of different tephras identified in any one profile, thus enhancing the temporal resolution of the overall chronology. Secondly, they are an essential step in extending the spatial coverage of tephras from distributions in the order of 10^2 km^2 to 10^4 km^2 to distributions of 10^5 km^2 to 10^6 km^2. Crucially, the presence of microscopic tephras is the key to effective correlations with the ice-core records, such as GRIP and GISP2 (Grönvold et al. 1995; Zielinski et al. 1995), and the resulting combination of high resolution temporal records (ice-cores) and precise, spatially extensive isochrons (tephra horizons).

Accurate identification and correlation of tephras will create a precise relative dating framework, but for the greatest utility 'numerical' dates have to be determined. In terms of developing the Icelandic Late Quaternary chronology, historical records and radiocarbon dating have been the dominant approaches more recently supplemented by correlation of Icelandic tephra to glass shards detected in the ice-cores (e.g. Thórarinsson 1967; Larsen and Thórarinsson 1977; Grönvold et al. 1995). Accurate historical dates have been attached to many tephras erupted within the last 900 years and sometimes these may be precise to the hour. For example, the tephra fall from the Hekla eruption of AD 1341 (H1341, Figure 8.4) began in southern Iceland from about 9 a.m. on the 19th of May (*Skálholtsannáll* quoted in Thórarinsson 1967). Ice-core dates have now been determined for two particularly important tephras formed around the Norse settlement of Iceland, and before the period of contemporaneous written records. The Landnám tephra from the Vatnaöldur fissure (Larsen 1984) was erupted about the time of the Norse colonization, traditionally dated on the basis of later saga writing to AD 874. The ice-core dates are AD 871 ± 2 (Grönvold et al. 1995) and AD 877 ± 4 (Zielinski et al. 1997). Shortly after the initial Norse colonization the largest basaltic fissure eruption of recorded history occurred on the Eldgjá fissure (Larsen 1979, 2000). A tentative date of the mid AD 930s based on an acid signal in the Crete ice core (Hammer et al. 1980) has been confirmed by tephrochronological correlation to the GISP2 core (Zielinski et al. 1995). These tephra layers are important for better understanding of the environmental impact of human colonization (Dugmore et al. 2000) and they have also been used to constrain a range of other environmental studies, such as recent glacial history (Figure 8.2).

Elsewhere in the world the historical record can also provide valuable dating evidence for constructing Late Quaternary tephrochronology, although this data is rather variable (Simkin and Siebert 1994). Northern European volcanism in the Massif Central and Eiffel districts is prehistoric in age, but contemporaneous accounts exist for Mediterranean tephrochronology, most notably the classic account of the AD 79 eruption of Vesuvius by Pliny the Younger (Radice 1969; Sigurdsson et al. 1985). Accurate contemporaneous human records of volcanic activity and tephra falls from the south-east Asian islands, Oceania, Kamchatka, the Western Hemisphere and east Africa generally post-date European colonization. Japanese historical data are notable because of the juxtaposition of a long local cultural record and widespread tephra production (Machida 1980; Simkin and Siebert 1994).

Radiocarbon dating is particularly important for the dating of Late Quaternary prehistoric tephras, both in Iceland and generally throughout the world (Simkin and Siebert 1994). The large spatial extent of tephras has two important implications for radiocarbon dating. Firstly, tephras may be traced to optimum sequences for undertaking radiocarbon dating, and secondly, precisely constrained multiple dates may be obtained on each tephra (Dugmore et al. 1995b, 2000). In addition, 'wiggle matching', using a sequence of high-precision dates and correlation to the calibration curve, can also be undertaken (Pilcher et al. 1995). Tephras occurring in both terrestrial and marine sediments can be used to help constrain the marine reservoir effect (e.g. Haflidason et al. 2000). Tephrochronology is also a route through which radiocarbon dating may be applied beyond the distribution of organic remains. The Katla-derived silicic tephra SILK YN has been dated in a peat bog using 22 separate radiocarbon dates, and the resulting error-weighted mean age of ca. AD 410 has been used to constrain the dating of moraines that lacked any organic deposits (Dugmore 1989; Dugmore et al. 2000; Larsen et al. 2001, and Figure 8.2). Outside Iceland and the North Atlantic region there are many similar examples of such tephrochronological applications from regions as diverse as Yellowstone in North America, the East African Rift Valley, central France, North Island New Zealand and the Dry Valleys of Antarctica that

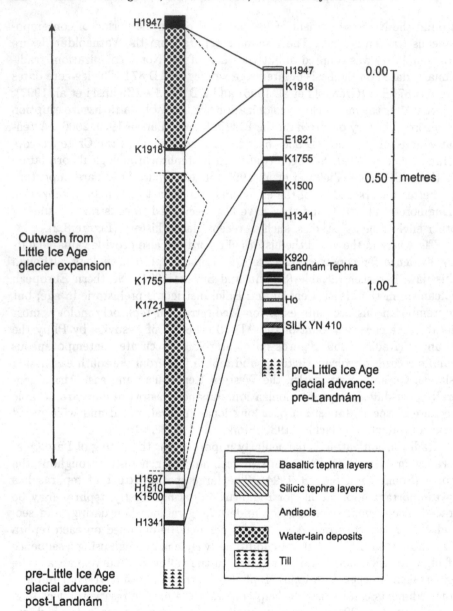

Fig. 8.4. Stratigraphic profiles from the foreland of Gígjökull, the principal outlet glacier flowing north from Eyjafjallajökull (Figure 8.3). Symbols follow the same conventions as Figure 8.3. The left-hand profile shows outwash sediments from Little Ice Age glacier advances intercalated with tephra and aeolian sediment, the right-hand profile is a reference section from within 200 m. The outwash sequences are bracketed between 16th-century tephras that include Hekla 1510, and the Hekla 1947 tephra (see also Figures 8.1 and 8.3).

illustrate wider-ranging contributions to constructing chronologies (Westgate et al. 1992; Sugden et al. 1995; Newnham and Lowe 1999; Lowe 2000).

8.3 Discussion

A key idea in 'crossing disciplines' is to develop applications of tephrochronology that turn the reporting/description/occurrence of tephra into an effective palaeoenvironmental tool with unique strengths. At its best tephrochronology should add insight at many different levels, and be greater than the sum of its constituent parts. Tephrochronology has been used to provide spot dates and limiting horizons at single sites, but there is considerable, and largely untapped potential for *3-D reconstructions* and the *spatial analysis* of patterns of change through time. Each tephra layer marks a surface at a *moment in time*; multiple tephra layers constrain the *passage of time*. In addition, the dispersal or reworking of tephra within sediments and contexts that are not contemporaneous with the formation of the tephra can be used to identify *processes* and *sediment pathways*. These include *temporary sediment stores* (e.g. fluvial terraces and englacial deposits), the *reworking of sediments* (e.g. glacio-fluvial sediments, archaeological sites) and *movements of sediments* (e.g. cryoturbation and bioturbation) both across landscapes and within profiles. These chronological concepts compliment geomorphological thinking elsewhere that identifies the differences between *landscape-forming* and *landscape-modifying* events (Manville 2001). The ways in which environmental records vary spatially through time are critical data, and essential constraints for effective modelling of environmental change. The 3-D reconstructions that can be created by mapping a series of tephra layers give uniquely detailed data on changing spatial patterns. This can be used to understand physical changes taking place in a landscape. For example, tephrochronology has been used to reconstruct the timing and spatial development of late Late Quaternary soil erosion in Iceland (Thórarinsson 1961; Dugmore and Buckland 1991; Dugmore and Erskine 1994). Crucially, tephrochronology can be used to evaluate landscape change in Iceland before and after human colonization and within periods of contrasting climates and cultures.

Once geomorphological research constrained by tephrochronology has determined the timing of *when* soil began to erode, the changing patterns of *where* soils eroded, and the probable physical processes of *how* the soil eroded can be tackled. The potential then exists to ask *why* the erosion took place. For this to succeed geomorphological data need to be constrained with sufficient precision for a meaningful integration with high-resolution climate data and, when the human impact comes into play, historical data on cultural systems and land management. The use of cultural chronologies in environmental analysis in Iceland requires both the identification of periods of time, as well as discrete events. These include periods such as the initial colonization (AD 870–930) and the Commonwealth period (AD 930–1262) and events including major epidemics, famines and cultural changes such as the establishment of a parliament and the introduction of different law codes, Christianity and colonial rule. Similarly, chronological control connecting climate records and geomorphology need to be able to resolve broad scale periods, such as the 'Little Ice Age' as well as distinct climate episodes such as cooler/wetter decades.

This is illustrated by the work of Simpson et al. (2001) in which detailed tephrochronological dating of landscape change facilitated a test of the hypothesis that the 'tragedy of the commons' accounts for historical rangeland erosion in southern Iceland. Their controversial conclusion, based on tephrochronology, geomorphological data, cultural records and biomass modelling, is that simple overstocking cannot explain the development of increased soil erosion during the historic period (Simpson et al. 2001). Effective management mechanisms were in place by the 13th century AD, and absolute biomass productivity has never appeared to be a constraint on historical stocking levels. It is probable that soil erosion in historical time increased orders of magnitude beyond pre-colonization levels because of a mismatch between the timing of rangeland grazing and the growing season, exacerbated by inadequate shepherding of livestock and limited fodder availability. This probably developed as a result of increasing climatic variability after the 14th century. Cultural systems were not able to track and manage unpredictable climate change, and as a result soil erosion spread and intensified (Arnalds 1999; Dugmore et al. 2000; Simpson et al. 2001). Crucially for this work, the spatial pattern and timing of soil erosion across entire landscapes had to be determined, and this was only possible through the use of tephrochronology. In this type of multidisciplinary approach, developing the conceptual ideas of *historical ecology* (Crumley 1994; Hunt and Kirch 1997), spatially referenced environmental data divided up on human time-scales (years and decades) is key. With these data, effective syntheses can be made of cultural and natural environmental records.

In the context of historical ecology, another aspect of tephrochronology is important. In addition to their function as marker horizons for the study of environmental change, tephra layers and the eruptions that created them may also be agents of environmental change. There have been numerous discussions about the possible environmental and cultural impacts of large-scale Icelandic tephra-forming eruptions in distal areas, many of which are currently unresolved (e.g. Baillie 1989, 1998; Pyle 1989; Buckland et al. 1997). Crucially, these ideas can be tested because of the widespread distribution of tephras in many different types of deposits that contain both environmental and cultural records.

In addition to the use of other dating techniques to ascribe 'numerical' ages to tephra marker horizons, these markers can be used to evaluate unrelated dating techniques such as lichenometry (Kirkbride and Dugmore 2001). Tephras have been used to independently date moraines and flood deposits, also subject to lichenometric dating. The tephras show that lichen ages produced by using established linear growth curves of extrapolated 20th-century data underestimate real ages. One implication is that the putative late 19th-century maximum extent of historical-age glaciation in Iceland could be an artefact of a flawed approach to dating (Kirkbride and Dugmore 2001). New approaches to lichenometry using non-linear growth curves that reflect the impacts of climate change on lichen growth rates have been stimulated as a result of the tephrochronological assessment of lichenometry, and first indications are that the new approach works well (Bradwell 2001).

8.4 Conclusions

1. Potentially, tephrochronology can be developed anywhere that identifiable tephra layers have been deposited and preserved.
2. The construction of a tephrochronology requires stratigraphic data that are spatially referenced, the mapping of tephra deposits, and the determination of characteristic signatures of the tephra that are typically defined by the chemical compositions of individual glass shards or minerals.
3. Relative dating control offered by individual tephras has to be connected to 'numerical' time scales. Using tephra isochrons, other chronological techniques (such as radiocarbon dating) may be combined, tested and applied either at optimum locations or extended to locations outside their normal spatial range.
4. Applications of tephrochronology in environmental studies should aim to turn the reporting/description/occurrence of tephra into an even more effective palaeoenvironmental tool with unique strengths. At its best tephrochronology should add critical insight at many different levels, and be greater than the sum of its constituent parts.

Acknowledgements

Aspects of this work have been supported by grants from the Leverhulme Trust, the Carnegie Trust for the Universities of Scotland, and the National Science Foundation of America. A number of key case studies have been facilitated by the NABO research co-operative. The chapter has benefited from the perceptive comments of Dr David Lowe.

References

Arnalds, O. (1999). The icelandic 'rofabard' soil erosion features. *Earth Surface Processes and Landforms*, **24**, 1–12.
Baillie, M. G. L. (1989). Hekla 3: how big was it? *Endeavour*, **13**, 78–81.

Baillie, M. G. L. (1998). Bronze age myths expose archaeological shortcomings? *Antiquity*, **72**, 425–427.

Boygle, J. (1998). A little goes a long way: discovery of a new mid-Holocene tephra in Sweden. *Boreas*, **27**, 195–199.

Bradwell, T. (2001). A new lichenometric dating curve for southeast Iceland. *Geografiska Annaler*, **83A**, 91–101.

Buckland, P. C., Dugmore, A. J. and Edwards, K. J. (1997). Bronze Age myths? Volcanic activity and human response in the Mediterranean and North Atlantic region. *Antiquity*, **72**, 424–432.

Connell, A. (1846). Analysis of the volcanic dust which fell in the Orkney Islands on the 2nd of September 1845. *Edinburgh New Philosophical Journal*, **40**, 217–220.

Crumley, C. (1994). Historical ecology: a multidimensional ecological orientation. In C. Crumley (ed.), *Historical ecology: cultural knowledge and changing landscapes*, School of American Research, Santa Fe.

Davies, S. M., Turney, C. S. M. and Lowe, J. J. (2001). Identification and significance of a visible, basalt-rich Vedde Ash layer in a Late-glacial sequence on the Isle of Skye, Inner Hebrides, Scotland. *Journal of Quaternary Science*, **16**, 99–105.

Dugmore, A. J. (1989). Tephrochronological studies of Holocene glacier fluctuations in south Iceland. In J. Oerlemans (ed.), *Glacier fluctuations and climatic change*, Kluwer Academic Publishers, Dordrecht, Netherlands.

Dugmore, A. J. and Buckland, P. C. (1991). Tephrochronology and late Holocene soil erosion in South Iceland. In J. Maizels and C. Caseldine (eds.), *Environmental change in Iceland*, Kluwer Academic Publishers, Dordrecht, Netherlands.

Dugmore, A. J. and Erskine, C. C. (1994). Local and regional patterns of soil erosion in southern Iceland. *Münchener Geographische Abhandlungen*, **13**, 63–79.

Dugmore, A. J., Newton, A. J., Edwards, K. J., Larsen, G., Blackford, J. J. and Cook, G. T. (1996). Long-distance marker horizons from small-scale eruptions: some British tephra deposits from the AD 1510 eruption of Hekla, Iceland. *Journal of Quaternary Science*, **11**, 511–516.

Dugmore, A. J., Newton, A. J. and Larsen, G. (1995a). Seven tephra isochrones in Scotland. *The Holocene*, **5**, 257–266.

Dugmore, A. J., Newton, A. J., Larsen, G. and Cook, G. T. (2000). Tephrochronology, environmental change and the Norse colonisation of Iceland. *Environmental Archaeology*, **5**, 21–34.

Dugmore, A. J., Shore, J. S., Cook, G. T., Newton, A. J., Edwards, K. J. and Larsen, G. (1995b). Radiocarbon dating tephra layers in Britain and Iceland. *Radiocarbon*, **37**, 379–388.

Einarsson, E. H., Larsen, G. and Thórarinsson, S. (1980). The Sólheimar tephra layer and the Katla eruption of c. 1357. *Acta Naturalia Islandica*, **28**, 1–24.

Froggatt, P. C. and Lowe, D. J. (1990). A review of late Quaternary silicic and some other tephra formations from New Zealand: their stratigraphy, nomenclature, distribution, volume, and age. *New Zealand Journal of Geology and Geophysics*, **33**, 88–99.

Grönvold, K., Óskarsson, N., Johnson, S. J., Clausen, H. B., Hammer, C. U., Bond, G. and Bard, E. (1995). Tephra layers from Iceland in the Greenland GRIP ice core correlated with oceanic and land based sediments. *Earth and Planetary Science Letters*, **135**, 149–155.

Haflidason, H., Eiríksson, J. and Kreveld, S. (2000). The tephrochronology of Iceland and the North Atlantic region during the middle and late Quaternary: a review. *Journal of Quaternary Science*, **15**, 3–22.

Hall, V. A. and Pilcher, J. R. (2002). Late Quaternary Icelandic tephras in Ireland and Great Britain: detection, characterisation and usefulness. *The Holocene*, **12**, 223–230.

Hammer, C. U., Clausen, H. B. and Dansgaard, W. (1980). Greenland icesheet evidence of postglacial volcanism and its climatic impact. *Nature*, **288**, 230–235.

Hunt, T. L. and Kirch, P. V. (1997). The historical ecology of Ofu island, American Samoa 3000 BP to the present. In P. V. Kirch and T. L. Hunt (eds.), *Historical ecology in the Pacific Islands*, Yale University Press, Newhaven, CT.

Jakobsson, S. P. (1979). Petrology of recent basalts of the Eastern Volcanic Zone, Iceland. *Acta Naturalia Islandica*, **26**, 1–103.

Jóhannesson, H., Flores, R. M. and Jónsson, J. (1981). A short account of the Holocene tephrochronology of the Snaefellsjökull central volcano, Western Iceland. *Jökull*, **31**, 23–30.

Kirkbride, M. P. and Dugmore, A. J. (2001). Can the late 'Littie Ice Age' glacial maximum in Iceland be dated by lichenometry? *Climatic Change*, **48**, 151–167.

Larsen, G. (1979). Um aldur Eldgjáhrauna (Tephrochronological dating of the Eldgjá lavas in south Iceland). *Náttúrufraedingurinn*, **49**, 1–26.

Larsen, G. (1981). Tephrochronology by microprobe glass analysis. In S. Self and R. S. J. Sparks (eds.), *Tephra studies*, D. Reidel, Dordrecht, 95–102.

Larsen, G. (1984). Recent volcanic history of the Veidivötn fissure swarm, south Iceland – an approach to volcanic risk assessment. *Journal of Volcanology and Geothermal Research*, **22**, 33–58.

Larsen, G. (2000). Holocene eruptions within the Katla volcanic system, south Iceland: characteristics and environmental impact. *Jökull*, **49**, 1–29.

Larsen, G., Dugmore, A. J. and Newton, A. J. (1999). Geochemistry of historical age silicic tephras in Iceland. *The Holocene*, **9**, 463–471.

Larsen, G., Gudmundsson, M. T. and Björnsson, H. (1998). Eight centuries of periodic volcanism at the center of the Icelandic hotspot revealed by glacier tephrostratigraphy. *Geology*, **26**, 943–946.

Larsen, G., Newton, A. J., Dugmore, A. J. and Vilmundardóttir, E. G. (2001). Geochemistry, dispersal, volumes and chronology of Holocene silicic tephra layers from the Katla volcanic system, Iceland. *Journal of Quaternary Science*, **16**, 119–132.

Larsen, G. and Thórarinsson, S. (1977). H4 and other acid Hekla tephra layers. *Jökull*, **27**, 28–46.

Lowe, D. J. (1988). Stratigraphy, age, composition, and correlation of late Quaternary tephras interbedded with organic sediments in Waikato lakes, North Island, New Zealand. *New Zealand Journal of Geology and Geophysics*, **31**, 125–165.

Lowe, D. J. (2000). Upbuilding pedogenesis in multisequal tephra-derived soils in the Waikato region. In J. A. Adams and A. K. Metherell (eds.), *Soil 2000: new horizons for a new century*. New Zealand Society of Soil Science, Australian and New Zealand Second Joint Soils Conference Volume 2: Oral Papers. 3–8 December 2000, Lincoln University, 183–184.

Lowe, D. J. and Hunt, J. B. (2001). A summary of terminology used in tephra-related studies. In E. T. Juvigné and J.-P. Raynal (eds.), *Tephras: chronology and archaeology*, Goudet, France, Les Dossiers de l'Archéo-Logis, **1**, 17–22.

Machida, H. (1980). Tephra and its implications with regard to the Japanese Quaternary period. In T. A. of Japanese Geographers (ed.), *Geography of Japan*, Teikoku-Shoin, Tokyo.

Mangerud, J., Furnes, H. and Johansen, J. (1986). A 9000 year old ash bed on the Faroe Islands. *Quaternary Research*, **26**, 262–265.

Mangerud, J., Lie, S. E., Furnes, H., Kristiansen, I. and Lomo, L. (1984). A Younger Dryas ash bed in western Norway, and its possible correlations with tephra in cores from the Norwegian Sea and the North Atlantic. *Quaternary Research*, **21**, 85–104.

Manville, V. (2001). Environmental impacts of large-scale explosive rhyolitic eruptions in the central North Island. In R. T. Smith (ed.), *Fieldtrip guides, Geological Society of New Zealand annual conference 2001*, Geological Society of New Zealand, Miscellaneous Publication, **110 B**, 1–19.

Mohn, H. (1877). Askeregnen den 29de-30te Marts 1875 (The tephra fall on 29–30 March 1875). *Norske Videnskabers Selskabs Forhandlinger (Royal Norwegian Society of Sciences and Letters)*, **10**, 1–12.

Newnham, R. M. and Lowe, D. J. (1999). Testing the synchroneity of pollen signals using tephrostratigraphy. *Global and Planetary Change*, **21**, 113–128.

Pilcher, J., Hall, V. A. and McCormac, F. G. (1995). Dates of Holocene Icelandic volcanic eruptions from tephra layers in Irish peats. *The Holocene*, **5**, 103–110.

Pilcher, J., Hall, V. A. and McCormac, F. G. (1996). An outline tephrochronology for the Holocene of the North of Ireland. *Journal of Quaternary Science*, **11**, 485–494.

Pyle, D. M. (1989). Ice core acidity peaks, retarded tree growth and putative eruptions. *Archaeometry*, **31**, 88–91.

Radice, B. (1969). *The letters of the Younger Pliny*. Penguin Books, Harmondsworth, England.

Saemundsson, K. (1991). Jardfraedi Kröflukerfisins (Geology of the Krafla volcanic system). Náttúra Mývatns, Reykjavík, Hid íslenska náttúrufraedifélag, 24–95.

Salmi, M. (1948). The Hekla ashfalls in Finland, AD 1947. *Suomen Geologinen Seura*, **21**, 87–96.

Shane, P. (2000). Tephrochronology: a New Zealand case study. *Earth Science Reviews*, **49**, 223–259.

Sigurdsson, H., Carey, S., Cornell, W. and Pescatore, T. (1985). The eruption of Vesuvius in AD 79. *National Geographic Research*, **1**, 332–387.

Simkin, T. and Siebert, L. (1994). *Volcanoes of the world*. Geoscience Press, Tucson, AZ, second edn.

Simpson, I. A., Dugmore, A. J., Thomson, A. and Vésteinsson, O. (2001). Crossing the thresholds: human ecology and historical patterns of landscape degradation. *Catena*, **42**, 175–192.

Sparks, R. S. J., Wilson, L. and Sigurdsson, H. (1981). The pyroclastic deposit of the 1875 eruption of Askja, Iceland. *Philosophical Transactions of the Royal Society of London Series A*, **299**, 241–273.

Sugden, D. E., Marchant, D. R., Potter, N., Souchez, R. A., Denton, G. H., Swisher, C. C. and Tison, J. L. (1995). Preservation of Miocene glacier ice in East Antarctica. *Nature*, **376**, 412–414.

Thórarinsson, S. (1944). Tefrokronologiska studier på Island. *Geografiska Annaler*, **26**, 1–217.

Thórarinsson, S. (1954). *The tephra fall from Hekla on March 29th 1947*. The eruption of Hekla 1947–1948, **II**. H. F. Leiftur, Reykjavík.

Thórarinsson, S. (1958). The Öræfajökull eruption of 1362. *Acta Naturalia Islandica*, **2**, 1–99.

Thórarinsson, S. (1961). Uppblástur á Íslandi íljósi öskulagarannsókna (Wind erosion in Iceland. A tephrochronological study). *Ársrit Skógrœktarfélags Íslands*, **1961**, 17–54.

Thórarinsson, S. (1967). *The eruptions of Hekla in historical times*. The eruption of Hekla 1947–1948, **I**. H. F. Leiftur, Reykjavík.

Thórarinsson, S. (1981). The application of tephrochronology in Iceland. In S. Self and R. S. J. Sparks (eds.), *Tephra studies*, D. Reidel, Dordrecht, 109–134.

Traill (1845). On the recent eruption of Hecla, and the volcanic shower in Orkney. *Proceedings of the Royal Society of Edinburgh*, **2**, 56–57.

Van den Bogaard, C. and Schminke, H.-U. (2002). Linking the North Atlantic to central Europe: a high resolution Holocene tephrochronological record from northern Germany. *Journal of Quaternary Science*, **17**, 3–20.

Walker, G. P. L. (1980). The Taupo plinian pumice: product of the most powerful known (ultraplinian) eruption? *Journal of Volcanology and Geothermal Research*, **8**, 69–84.

Wastegård, S., Turney, C. S. M., Lowe, J. J. and Roberts, S. J. (2000a). The Vedde Ash in NW Europe: distribution and geochemistry. *Boreas*, **29**, 72–78.

Wastegård, S., Wohlfarth, B., Subetto, D. A. and Sapelko, T. V. (2000b). Extending the known distribution of the Younger Dryas Vedde Ash into north western Russia. *Journal of Quaternary Science*, **15**, 581–586.

Westgate, J. A. and Gorton, M. P. (1981). Correlation techniques in tephrastudies. In S. Self and R. S. J. Sparks (eds.), *Tephra studies*, D. Reidel, Dordrecht, 73–94.

Westgate, J. A., Walter, R. C. and Naeser, N. (eds.) (1992). *Tephrochronology: stratigraphic applications of tephra*, vol. 13 and 14. Special Volumes of Quaternary International.

Zielinski, G. A., Germani, M. S., Larsen, G., Baillie, M. G. L., Whitlow, S., Twickler, M. S. and Taylor, K. (1995). Evidence of the Eldgjá (Iceland) eruption in the GISP2 Greenland ice core: relationship to eruption processes and climatic conditions in the tenth century. *The Holocene*, **5**, 129–140.

Zielinski, G. A., Mayewski, P. A., Meeker, L. D., Grönvald, K., Germani, M. S., Whittlow, S., Twicker, M. S. and Taylor, K. (1997). Volcanic aerosol records and tephrochronology of the Summit, Greenland, ice cores. *Journal of Geophysical Research*, **102**, 26625–26640.

9

Constructing Chronologies of Sea-Level Change from Salt-Marsh Sediments

Robin J. Edwards

Summary. This chapter reviews recent research into the chronology of sea-level change. Since sea-level is related to global temperatures, an understanding of sea-level change has become particularly important over recent years and is offering insights into a range of issues including ice sheet distribution, past environmental change and coastal management. In all these areas, an understanding of the timing as well as the nature of sea-level change is important and so chronometric methods as well as stratigraphic ones are typically employed. This chapter reviews the nature of the data that provide information about sea-level change and highlights some of the challenges facing those who wish to establish chronologies. The challenges have parallels with those in archaeological research (compare Chapters 1, 2 and 4), but as yet there has not been much cross-fertilization of methods between the two disciplines. This chapter offers some suggestions for the most pressing problems that still need formal tools and provides a case study using wiggle-matching that illustrates the kinds of improvements that researchers of sea level change might expect if such methods were adopted more widely.

9.1 Introduction

Geologically-based sea-level research spans a wide range of spatial and temporal scales, from 'global' glacial/interglacial variations to local, multi-decadal, decimetre perturbations. Consequently, an accurate knowledge of sea-level change can contribute to our understanding of such diverse issues as ice sheet distribution, Earth rheology, ocean–atmosphere interaction, palaeoenvironmental change and coastal management. Irrespective of the ultimate research objective, the establishment of a reliable chronology is fundamental to all sea-level studies.

The most recent period of Earth history (called the 'Quaternary') began around 2.5 million years ago and was characterized by a series of perhaps 50 oscillations in global climate. These climate changes involved the growth of large terrestrial ice sheets during cold, glacial periods, and their subsequent decay as temperatures warmed during inter-glacials. During these 'ice ages', sea levels fell as large volumes of water were removed from the oceans and incorporated into the expanding ice sheets. When global temperatures began to increase once more, such as happened toward the end of the last glacial period around 19,000 years ago, this water was released back into the oceans and sea levels rose. By the start of the current warm period around 10,000 years ago (called the 'Holocene'), many coastal areas were being inundated by the rising oceans. Not all coastlines witnessed the same change in sea level however. Large areas of land that had been depressed under the weight of these ice sheets were now rebounding and getting higher themselves. Consequently, the actual change evident at any point around the coast was a combination of vertical movements in both sea and land levels. For this reason, sea-level researchers who measure sea-level change relative to the land, tend to refer to variations in 'relative sea-level'.

Relative sea-level changes during the Holocene have been measured in a variety of ways, reflecting the type of coastline under investigation. On rocky shores for example, encrustations of marine organisms (e.g. barnacles) and morphological features (e.g. wave-cut notches) have been employed as sea-level indicators (e.g. Pirazzoli 1986; Laborel et al. 1994; Mastronuzzi and Sansò 2002). Similarly, sea-level data from tropical regions are often derived from measuring the morphology of coral reefs (e.g. Fairbanks 1989). A full consideration of these diverse methods and their limitations is beyond the scope of this review, and the interested reader is referred to the more general introduction provided by Pirazzoli (1996). Instead, this chapter concentrates on the chronological issues that surround the construction of relative sea-level records from sedimentary coastlines. The methods presented here, and their derivatives, have been used in a wide range of locations including salt-marshes in Japan (Sawai et al. 2002); mangrove systems in Bangladesh (Shahidul Islam and Tooley 1999), and isolation basins in Greenland (Long and Roberts 2002). For the purposes of this discussion, selected reconstructions from the UK and USA are presented to highlight key issues relating to the construction of chronologies from these types of environment. More specifically, attention is focused on the development of chronologies from the late Holocene period (covering the last 3000 years or so) since these present the greatest challenges, both in terms of construction (resolution) and application (precision and correlation).

This chapter begins with an overview of the methods used to extract information on relative sea-level changes from coastal sedimentary sequences, concentrating on the generation of data from the last two to three millennia. This is not intended as a comprehensive review of the diverse nature of sea-level research, but rather presents selected examples to illustrate some key issues related to establishing chronologies from these environments and this time period. These issues include the uneven spatial and temporal distribution of dateable material, the relative magnitude of age uncertainties, and problems associated with age interpolation. Some potential solutions to these problems are then discussed including the use of composite chronologies, chronohorizons, short-lived radionuclides (e.g. ^{210}Pb) and archaeological evidence to increase the frequency and distribution of age markers in sedimentary sequences. Finally, a chronology of wiggle-matched AMS radiocarbon dates is presented to illustrate the type of high-precision age control that can be obtained from organic-rich salt-marsh deposits.

9.2 Sea-Level Records from Sedimentary Coastlines

The interchange between land and sea may be recorded in coastal sediments as changes between organic-rich accumulations of freshwater peat (terrestrial), and organic-poor deposits of sand, silt and clay (marine). The investigation of buried peat beds has a long history in the UK, due in large part to the pioneering work of Godwin (e.g. Godwin 1940, 1945; Godwin and Godwin 1940; Godwin et al. 1958). Care must be taken when interpreting these deposits in terms of sea-level change since terrestrial peat can form at a range of elevations above sea-level and so is not always a reliable sea-level indicator. This problem can be avoided by focusing on the transitions between freshwater and marine environments, which tend to occur at or around mean high water spring tides (Shennan 1986). By quantifying the vertical relationship between the tidal frame and a sediment body at the time of its deposition (called the indicative meaning; van de Plassche 1982, 1986; Shennan 1986), it is possible to produce more reliable reconstructions of relative sea-level. Examining the salt-marsh sediments that form during these transitional phases can further increase the precision of these reconstructions.

Salt-marshes exist at the interface between land and sea, and their vertical range is closely related to the tidal frame (Pethick 1980, 1981; Allen 1990a,b,c). Their usefulness as sea-level indicators is enhanced by a pronounced vertical zonation of flora and fauna, related to elevation above mean tidal level (Chapman 1960; Scott and Medioli 1980; Gray 1992). This vertical differentiation between salt-marsh subzones and adjacent freshwater or mudflat environments can be identified in fossil deposits, enabling more precise determination of their indicative meaning.

Two distinct approaches to sea-level reconstruction have been employed in sedimentary environments of the type outlined above. The first method considers vertical movements in relative sea-level and attempts to fix its altitude in time and space (age–altitude analysis). The second approach examines lateral shifts in depositional environment that occur in response to relative sea-level change and sedimentary processes (tendency and marsh palaeoenvironmental analysis). Irrespective of the method employed however, the construction of a precise and accurate age–depth relationship or accumulation history is central to the reliable reconstruction of past sea-level change. The most common means of providing age information is radiocarbon dating of organic material within these sedimentary sequences.

This chapter presents an overview of these contrasting approaches to sea-level reconstruction, focusing on the requirements and limitations associated with reliably establishing radiocarbon-based chronologies for the late Holocene period. It concludes by examining some potential ways that existing chronological frameworks may be improved.

9.3 Reconstructing Vertical Relative Sea-Level Movements

Records of Holocene changes in relative sea-level from around the UK have been constructed by studying variations in coastal sedimentary sequences and associated biological components contained within them (e.g. Shennan 1982; Long and Tooley 1995; Shennan et al. 1995). A multi-proxy approach employing radiocarbon-dated lithostratigraphic and biostratigraphic sea-level indicators is routinely used to establish a series of sea-level index points that fix former positions of relative sea-level in terms of age and altitude. Errors associated with the precise determination of this age and altitude information require these data to be plotted as a band of sea-level change, representing the generalized course of variations in relative sea-level throughout the study period (Shennan 1986). The application of this methodology has provided important information on a range of issues, such as UK crustal movements, derived from comparison of regional relative sea-level histories from around the coastline of the British Isles (Shennan and Horton 2002).

The limitations associated with error terms become increasingly problematic as the temporal scale of the investigation is reduced. For example, when reconstructing sea-level changes during the last two or three millennia, the error band becomes of comparable magnitude to the sea-level variations of interest. A complication in many NW European marshes is the relative paucity of organic sediments required for radiocarbon dating. For example, a dramatic reduction in organic accumulation after c. 1000 BC can be observed in a number of coastal records from the UK (e.g. Devoy 1979; Allen 1991; Shennan 1994; Long et al. 2000), and this has frustrated attempts to resolve recent sea-level changes. The consequence is that many studies lack any data from the last two or three millennia, and resulting chronologies of change are often in part an artefact of the temporal distribution of data points.

9.4 Reconstructing Lateral Changes in Palaeoenvironment

An alternative approach in sea-level research has been to reconstruct lateral shifts in terrestrial and marine environments that reflect a combination of changes in marsh accumulation and relative sea-level movements. This type of approach commonly involves the construction of marsh palaeoenvironmental curves that plot changes in water depth (submergence and emergence) associated with the variations in depositional environment. The marsh palaeoenvironmental curve methodology has played a central role in high-resolution studies from North American salt-marshes seeking to evaluate the relationship between climate and sea-level changes (e.g. Varekamp et al. 1992; van de Plassche 2000).

The salt-marshes of the American Eastern Seaboard, particularly those around Maine, Connecticut and Massachusetts, are ideally suited to sea-level research as they possess highly organic sedimentary sequences that permit a high-resolution, radiocarbon chronology to be constructed. The marsh palaeoenvironmental curve approach is most suited to the organic-rich marshes of North America since accumulation is predominantly controlled by the growth of salt-marsh plant communities that are closely linked to elevation relative to the tidal frame, and hence contain a strong sea-level signature. Conversely, thick organic sequences are prone to compaction, either under their own weight (autocompaction), or as a result of loading from overlying sediments. A sea-level index point approach attempting to fix the vertical position of past relative sea-level will be in error unless this compaction effect can be quantified, or un-compacted (basal) samples are used. Whilst a number of attempts have been made to quantify compaction in the salt-marshes of America (e.g. Pizzuto and Schwendt 1997), where deposits may be displaced by as much as one metre (Gehrels 1999), at present there is no established methodology to deal with this spatially variable, multi-faceted phenomenon.

9.5 Key Dating Limitations

The two methodologies outlined above, or variants of them, are responsible for the majority of sea-level data produced in temperate latitudes where salt-marshes are present. Their chronologies are usually based on radiocarbon-dated samples and as such share a number of characteristic limitations. Potential sources of error associated with radiocarbon dating are reviewed by Mook and van de Plassche (1986), Aitken (1990), and Switsur (1994), and will not be considered in detail here. Instead, attention will be focused on the application of these data in the context of sea-level research.

9.5.1 Available Carbon

Conventional radiocarbon dating requires comparatively large volumes of organic material ($\geq 90\,g$) that can equate to a sample thickness of 5 cm or more from narrow bore coring devices sometimes used in sea-level investigation. Thick samples have correspondingly longer periods of formation, and therefore age estimates represent an integration of the time elapsed during sediment deposition. In high marsh contexts, sedimentation rates are usually low (Allen

1990c), and therefore substantial age errors may be introduced in this way. In addition the greater volume of material required demands collection of secondary cores for microfossil analysis, and thereby introduces errors associated with cross-correlation. The use of larger diameter, mechanical coring devices can reduce this problem, although the logistical difficulties of transport and deployment in remote areas with restricted access and waterlogged substrates can be prohibitive.

The development of accelerator mass spectrometry (AMS) radiocarbon dating allows much smaller quantities of carbon to be used ($<1\,g$). This reduces the errors outlined above and expands the range of environments from which dated material can be recovered. Care must still be taken to avoid the incorporation of 'old carbon' that may have been washed into a sample from elsewhere, since this will produce erroneous age estimates. Such contamination is likely to be most pronounced when attempting to date bulk samples with low organic content. AMS dating is best applied to plant macrofossils, especially those that are *in situ* and have quantified vertical relationships to the marsh surface (e.g. van de Plassche 2000). Whilst macrofossils are abundant in some US marshes where they form the core of dating programmes, preservation in British marshes is commonly poor and further hinders the establishment of reliable dated horizons.

Irrespective of the technique used, samples with sufficient organic material for radiocarbon dating are not present throughout the sedimentary sequences of minerogenic UK and NW European marshes. As a consequence, horizons with established ages will not be uniformly distributed through a sediment core and the resulting chronology will be of variable precision and in part influenced by data availability. Figure 9.1 shows an example age–altitude diagram of sea-level index points from the Solent in southern Britain, established by conventional radiocarbon dating of lithostratigraphic contacts between terrestrial and marine sediments. This shows that whilst there are seven dates from the last 5500 years (an average of one date per *c*. 800 years), due to the uneven distribution of data the real resolution is lower than this, particularly during the last 2000 years.

A note regarding financial constraints is also pertinent at this point since recent studies have demonstrated that differing age estimates can be returned from a single sample depending on the organic fraction used or the type of sample material selected (e.g. Turney et al. 2000). Replicate dating of samples from the same or similar contexts will clearly increase confidence in any resulting chronology by permitting some form of cross-validation to be performed. In sea-level research, dating is one of the principal financial restrictions and it is unclear how funding bodies will respond to a doubling in requested support. Confidence in age estimates will be increased further when alternative methods of dating are used to provide independent assessments, thereby avoiding the chance of compounding systematic errors (Sutherland 1982).

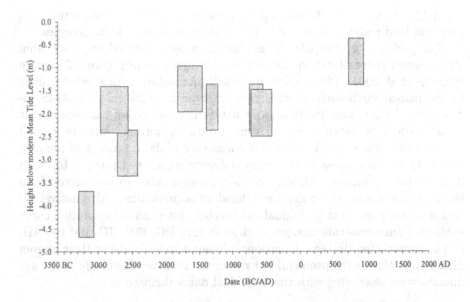

Fig. 9.1. Traditional sea-level index points from the Solent region for the last 5500 years, based on conventionally radiocarbon-dated lithostratigraphic contacts between terrestrial and marine sediments (2σ errors). The diagram shows the deviation of mean tide level from its present altitude.

9.5.2 Age Uncertainties and Interpolation

Variations in atmospheric ^{14}C activity and the statistical nature of radioactive decay mean that the calibration of an individual ^{14}C age estimate will produce a range of calendar ages. This age range can span up to 400 years where the interval of interest coincides with a period when the ^{14}C calibration curve exhibits a 'plateau' (e.g. the 'Hallstatt plateau' *c.* 2450 ^{14}C year BP). The variable nature of these age uncertainties can introduce considerable scatter in age–depth diagrams and complicate the construction of reliable accumulation curves. This is most pronounced if 2σ error ranges are used.

Ultimately, a simple substitution of age for depth is required and this will involve some form of interpolation between data points. At one extreme, this may take the form of a single, linear interpolation through the entire set of dates, whilst at the other it may involve multiple interpolations between individual data points. Neither of these approaches is without problems. In the case of the linear interpolation, a number of short-period but significant variations in accumulation rate may be masked beneath a single, generalized value. Conversely, visual interpolation of a 'best-fit curve' is obviously subjective and will tend to give added weight to the data points with small age errors, corresponding to the steep sections of the ^{14}C calibration curve. Whilst attempts have been made to model salt-marsh sediment accumulation patterns (e.g.

Allen 1990c), the lack of data regarding variables such as compaction and sediment load restricts their use in formulating depositional chronologies.

The problem of interpolation is illustrated by the dated sequence from Pattagansett River Marsh in Connecticut, USA, shown in Figure 9.2 (van de Plassche et al. 2001). Plant macrofossils with quantified vertical relationships to the palaeomarsh surface were collected from a single core for AMS radiocarbon dating. The highly organic nature of these sediments meant that samples could be taken every c. 10 cm, producing an age estimate around every 80 calendar years. Despite this abundance of data, a range of possible accumulation curves may be constructed depending upon the degree to which data are (over-) interpreted. For example, a single rate can be inferred from these data, following the generalized 'band' of accumulation. Alternatively, if added weight is given to individual radiocarbon dates, an accumulation curve exhibiting numerous rate changes, such as at 2225 BC, 1350 BC, and 100 AD, can be constructed. In these situations it is not a lack of data or their uneven distribution that is the problem, but rather the relative magnitude of the age uncertainties associated with the individual dates themselves.

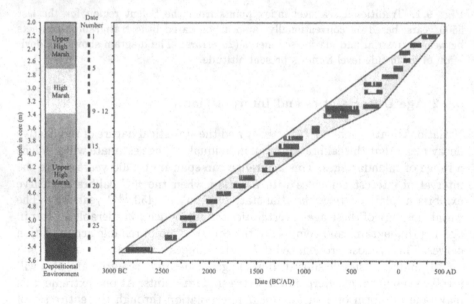

Fig. 9.2. Stratigraphic column and associated individually calibrated AMS radiocarbon dates (1σ) from Pattagansett River Marsh, Connecticut, USA (adapted from van de Plassche et al. 2001).

9.6 Some Potential Solutions

The problems with radiocarbon-based chronologies presented above fundamentally limit the quality and resolution of sea-level reconstructions. Improved age control may be possible by supplementing radiocarbon-dated evidence with age estimates provided by other chronostratigraphic tools to produce composite chronologies. Alternatively, a statistical treatment of sequences of radiocarbon dates can be attempted to increase the precision of the resulting calendar ages.

9.6.1 Composite Chronologies

The interplay between organic 'peat' accumulations and inorganic sand, silt or clay sediments has been discussed above. Despite the central role such intercalated sequences have in sea-level reconstruction, chronologies are commonly constructed from the organic component alone. The development of complementary chronologies derived from organic-poor sediments will significantly improve the temporal precision and resolution of sea-level records. Since radiocarbon techniques cannot be used to date such sediments, it will be necessary to construct 'composite chronologies' that employ different dating methods and seek to link these separate sets of data together to produce a coherent record. The use of alternative radiometric methods such as short-lived radionuclides has the potential to produce a more evenly distributed, and higher frequency, suite of dated points.

Short-Lived Radionuclides

A variety of short-lived radionuclides have been used to investigate salt-marsh accretion in the UK. Perhaps the most commonly used radionuclide in a sea-level context is ^{210}Pb, but since this can only provide useful information for the last 100 years or so it is of limited utility. Consequently, a considerable gap exists between the demise of organic accumulation suitable for radiocarbon dating and the deposition of sediments that can be analysed by other radiometric means. Even when the duration of this gap is short, the choice of method to join two data sets with greatly differing error terms is not always straightforward.

An example of a composite chronology is provided by the recent paper of Gehrels et al. (2002), in which six AMS radiocarbon dates are used in conjunction with ^{210}Pb / ^{137}Cs analysis and pollen data. Figure 9.3 presents the radiometric data and illustrates that, whilst clearly improving the age control in the last 400 years of the record, joining the two sets together presents a challenge.

The approach adopted by Gehrels et al. (2002) is to split the data into its component parts. First a (subjective) best-fit interpolation of the ^{14}C data is performed, and this is then joined to the ^{210}Pb chronology (black line in

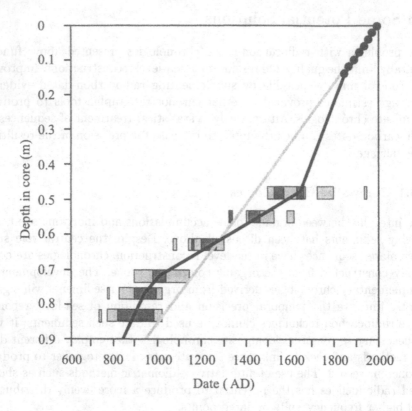

Fig. 9.3. An example of a composite chronology employing age estimates provided by ^{210}Pb analysis (circles) and radiocarbon dating (squares), from Wells, Gulf of Maine, USA (adapted from Gehrels et al. 2002). Dark grey boxes indicate the 1σ errors whilst the pale grey boxes indicate the 2σ range. The lines represent two possible methods of interpolation.

Figure 9.3). This has the effect of inducing a step in the accumulation rate at the junction between the two data sets. In this case, support for the chronology is fortuitously provided by a well-placed pollen chronohorizon, but there is no *a priori* reason to suggest that this will be possible everywhere. A danger exists therefore, that apparent changes in accumulation rate will be introduced simply by switching between sources of age data. These rate changes may then be transmitted into the resultant sea-level records.

An alternative would be to treat the data set as a single entity by establishing an accumulation rate on the basis of the ^{210}Pb record and then extrapolating this down-core with some key radiocarbon dates to assess the constancy of accumulation through time (pale grey line in Figure 9.3). This type of approach rests upon the assumption that, during the period under investigation, the marsh has generally accumulated at a constant rate. As such it is prone to considerable errors where relatively brief changes in depositional environment and consequently sedimentation rate have occurred.

Discrepancies between short-term accumulation rates (decades) and longer-term marsh accretion (centuries) are well known. For example, Cundy and Croudace (1996) used a range of short-lived radionuclides to calculate rates of salt-marsh accretion in the Solent region. Their results indicated that since 1900 AD, most sites had been accreting at 4 to 5 mm per year. Long et al. (1999) note a considerable acceleration in sedimentation rate from 1.14 to 7.17 mm per year associated with the colonization of the salt-marsh by common cord-grass (*Spartina anglica*). This is in good agreement with a value of 7.2 mm per year presented for a neighbouring marsh in Poole Harbour by Cundy and Croudace (1996). Clearly, extrapolation of a short-time series in a situation where the current marsh is adjusting to a new sedimentary regime will produce erroneous age–depth relationships when projected down-core. Furthermore, the inorganic sequences of the UK are associated with multiple hiatuses caused by erosion or non-deposition, perhaps during phases of reduced relative sea-level rise or relative sea-level fall (Edwards 2001). This 'missing time' warns against adoption of a single-core approach such as is employed in high-resolution sea-level investigation of the contrastingly organic-rich American marshes. Instead, UK records must be pieced together from multiple cores, exploiting the fact that the sensitivity of individual locations will vary through time with changing coastal configuration.

Luminescence Dating

Recent research has attempted to employ luminescence dating techniques to provide age estimates from minerogenic sediments in coastal contexts (e.g. Bailiff and Tooley 2000; Clarke and Rendell 2000). Attention has focused on the newer techniques of infra-red-stimulated luminescence and optically stimulated luminescence. Currently a number of problems are associated with the application of luminescence dating in water-lain environments, including incomplete zeroing of sample grains before burial ('poorly-bleached'), weathering of minerals and variations in the water content of sediments. Attention is being focused on ways to address these complications and some promising early results (e.g. Bailiff and Tooley 2000) suggest that this technique may become increasingly useful in the future.

In addition to directly dating sediments, the integration of sea-level research with coastal archaeological investigations provides the opportunity to establish a depositional framework on the basis of cultural evidence contained within the sequences. Traditional thermoluminescence dating of pottery for example, could prove particularly useful in constraining chronologies during the last 1000 years or so, when the resulting error terms are similar to or smaller in magnitude than those of radiocarbon dating (Aitken 1990). The context of the finds is of paramount importance in determining their reliabil-

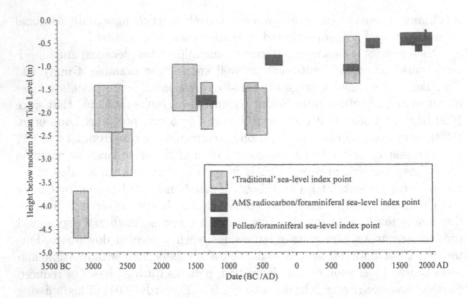

Fig. 9.4. Sea-level index points from the Solent region for the last 5500 years, including those established from AMS radiocarbon dates (2σ) and pollen chronohorizons. The diagram shows the deviation of mean tide level from its present altitude.

ity as age indicators, and consequently the involvement of the archaeological community will be necessary if this type of approach is to be successful.

Chronohorizons

Another approach is to supplement radiocarbon-dated chronologies with markers of known age contained within the sediment. One method employed in sea-level research is that of pollen chronohorizons, which uses a rise in the percentage abundance of spores from key species introduced at known times to date sediments. Long et al. (1999) and Edwards (2001) used the historically dated rise in common cord-grass (*Spartina angelica*) and pine (*Pinus*) pollen to provide age estimates (*c.* 1890–1900 AD and *c.* 1700–1800 AD respectively) for a sequence in Poole Harbour. Figure 9.4 shows the conventionally dated sea-level index points presented in Figure 9.1 plotted with additional index points derived from the two pollen chronohorizons and AMS radiocarbon dates. This demonstrates how the combination of improvements in radiocarbon dating and the integration of floral chronohorizons can increase the temporal resolution of a study, particularly for the last 2000 years. It also serves to highlight that the limited distribution of these indicators still results in considerable gaps in the record.

Other examples of age markers include the use of chemical or pollution indicators that can be related to events or processes of a known age. In the Severn Estuary for example, Allen (1987a,b, 1988) and Allen and Rae (1986) have utilized the metal and coal dust content of sediments to determine their ages by comparison with the documented increase of industry and associated pollution. The efficacy of this approach will be site specific, depending on the frequency, range and distribution of known inputs and the reliability of their associated ages.

In a similar way, volcanic glass (tephra) related to eruptions of known date may also be used to provide chronohorizons. Tephrochronology has been successfully applied to the dating of horizons in a range of environments from lakes to peat bogs. The applicability of these techniques will be geographically specific as ash layers are spatially restricted (Lowe and Walker 1997).

Archaeology and Tendency Analysis

Sea-level tendencies consider the timing of increases or decreases in marine influence at a site (Shennan et al. 1983). Since they do not require precise determination of altitude the approach enables a variety of dated sea-level indicators to be used in combination, without the need to quantify their indicative meanings. This expands the range of material that can be used to construct a temporal framework for sea-level change.

One classic application of tendency analysis is the Fenland chronology of Shennan (1986), derived from 112 radiocarbon-dated sea-level index points. This analysis identified a series of periods dominated by positive (Wash) and negative (Fenland) tendencies. In a contrasting approach, Waller (1994) examined the timing and extent of marsh expansion and marine incursion using palaeogeographic maps, and identified three distinct phases of change during the same period using 83 radiocarbon-dated sea-level index points. Despite working in the same area with similar data, there is considerable disagreement between the Wash/Fenland chronology of Shennan (1986, 1994) and the palaeogeographic interpretation of Waller (1994). The transformation of two-dimensional sea-level data into three-dimensional palaeogeographic maps is still in its infancy, but is now being used to investigate issues such as tidal range change (Shennan et al. 2000). Ultimately, it is this spatial expression of vertical changes in relative sea-level that is of paramount importance when reconstructing patterns of coastal evolution and inferring their relationship to human occupation and land use.

The discrepancies evident in the Fenland record may be related to a number of limitations associated with the use of tendency analysis. Tendencies are applied in a hierarchical fashion, constructing dominant or regional tendencies from a number of local ones. This assumes that regional processes, such as changes in sea-level, will act to produce widespread local tendencies of similar sense (Shennan et al. 1983). In reality, the local tendencies may be strongly influenced by other processes, such as changes in sediment supply, which will complicate production of a dominant tendency (Long 1992). Secondly, delimiting the period of a tendency is strongly affected by the temporal

202 Robin J. Edwards

distribution of data points. Consequently, tendency records require large numbers of dated sea-level indicators, and their chronology is strongly influenced by data distribution and sampling strategy (Long 1992).

It should be noted at this point that the bulk of the data used to produce the Fenland palaeogeographic maps and tendencies was palaeoecological in nature. One potential avenue worthy of further exploration is the use of cultural data from archaeological investigations to complement such palaeoenvironmental reconstruction. Rare examples can be found in the literature, such as the use of sea banks to delimit sedimentation within the Severn Estuary (Allen 1991). In general however, there has been a tendency for sea-level investigations to avoid areas that have experienced significant human modification and the inevitable 'contamination' of the record this would bring.

Recent years have witnessed a growth in coastal archaeology, driven in part by the growing realization that it represents a poorly understood and non-renewable resource that is under threat in regions of eroding coastline (Fulford et al. 1997; Pye and Allen 2000). These investigations commonly include a palaeoenvironmental component, aimed at providing a context for the archaeological data. Sampling strategies are usually driven by the archaeology and often take the form of 'mud in a bag' collected as part of the excavation process. In these instances, the analyst's first contact with the material commonly occurs when removing it from bubble-wrap in the laboratory. Interdisciplinary involvement early in the planning stage would prove beneficial to both communities, ensuring that the palaeoenvironmental data collected are of maximum utility. This does not necessitate the collection of more samples, but rather the development of strategies and field protocols that will generate data in a format that is useful for a variety of purposes. In a sea-level context, simple examples of this would be the collection of material for dating from the upper contact of a peat body rather than within the peat itself, and ensuring that the altitude of the sediment surface is recorded in addition to sampling depth. Ideally, both communities should be involved in the collection of samples and not just their subsequent processing.

9.6.2 Wiggle-Match Dating (WMD)

The discussion so far has concentrated on improving chronologies by employing additional forms of age control to supplement existing radiocarbon data. Where large numbers of radiocarbon dates are available however, improvements can be made by changing the way in which these data are used. An example of this is wiggle-matching, in which the stratigraphic relationship between radiocarbon dates from a single core, such as Pattagansett River Marsh (Figure 9.2), is used to refine the calibration process.

The wiggle-matching approach relies on the fact that atmospheric ^{14}C activity has varied through time and, as a consequence, the calibration curve relating ^{14}C ages to calendar dates exhibits numerous 'wiggles' (Figure 9.5 – data from Stuiver et al. 1998). These wiggles are problematic when considering individual dates in isolation, since they can produce multiple calendar dates for a single radiocarbon age. However, this same variability can be used to improve the precision of radiocarbon age estimates when multiple samples with known vertical relationships are considered in groups. In this instance, the age of adjacent samples will be related to their vertical separation by the accumulation rate of the sedimentary sequence. Consequently, by assigning an appropriate accumulation rate, a floating chronology can be developed. This chronology can be anchored in time by using distinctive 'wiggles' in the calibration curve to produce a precise suite of calendar ages. In practice, the selection of the most appropriate accumulation rate is achieved via a computer package such as $Cal25$ (van der Plicht 1993), which performs a linear stretch of the calendar date axis to achieve a 'best-fit' defined in terms of some statistical measure. In the case of the $Cal25$ program, this best-fit is based on the squared sum of the difference between the data points and the calibration curve.

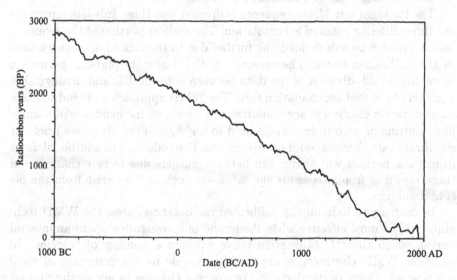

Fig. 9.5. Variations in the ^{14}C calibration curve during the last 3000 years (data from Stuiver et al. 1998).

Van de Plassche et al. (2001) applied the WMD approach to suites of AMS radiocarbon dates derived from salt-marsh peat cores, and the results of one of these wiggle-matches is presented here. The series of dates is taken from the Pattagansett River Marsh core presented in Figure 9.2. The starting point is to consider the simplest accumulation history for the sediment core and wiggle-match the entire suite of dates to produce a single accumulation rate. The resulting set of calendar ages differs from a simple linear interpolation in that it takes into account variations in atmospheric ^{14}C activity. Where this uniformitarian assumption is invalid, the sequence of dates progressively 'drifts' away from the calibration curve, indicating a faster or slower rate of

accumulation. Where this deviation occurs and associated lithological or biological indicators suggest environmental changes that may have influenced the
accumulation rate, the sequence of dates is sub-divided and wiggle-matched
separately. Where this sub-division occurs, individual samples are used as
'tie-dates' to link the suites of dates together and ensure stratigraphic relationships are maintained. In this way, the accumulation history of the sequence
is refined by successive iterations.

Figure 9.6 shows the results of this wiggle-match approach as it is applied
to the Pattagansett Marsh core. Wiggle-match dating was performed using
the Groningen radiocarbon calibration program (*Cal25*) (van der Plicht 1993),
which had previously been used in the dating of raised peat bog deposits (e.g.
van Geel and Mook 1989; Kilian et al. 1995). The *Cal25* program does not
provide an indication of errors for the resultant WMD calendar ages, and here
an arbitrary value of 50 years is assigned to demonstrate the existence of an
uncertainty, after van de Plassche et al. (2001).

The Pattagansett Marsh sequence is divided into three sub-sets representing three differing rates of accumulation. The earliest portion of the sequence
cannot reliably be sub-divided any further due to the lack of distinct features
in the calibration curve. The presence of the 'Hallstatt plateau' permits a
more precise sub-division of the data between 800–400 BC and distinguishes
a period of reduced accumulation rate. The WMD approach will tend to mask
short duration changes in accumulation rate although the existence of a single
date plotting as an outlier may be used to highlight intervals where brief but
significant rate changes could have occurred. The collection of additional dates
from these periods will distinguish between outliers due to rate changes and
those resulting from erroneous age estimates such as may arise from sample
contamination.

In contrast to individually calibrated radiocarbon dates, the WMD technique will be most effective when the period of investigation spans an interval
during which the ^{14}C calibration curve exhibits a number of 'wiggles'. In
addition, WMD chronologies are readily testable by the collection of more
radiocarbon dates, particularly in areas where changes in accumulation rate
are inferred. For example, by increasing the sampling frequency during intervals such as the 'Hallstatt plateau', the timing of changes can be more
precisely and accurately matched. This is also in direct contrast to the use
of individually calibrated dates, where the collection of more data frequently
serves to increase the observed scatter and complicate the reconstruction of
accumulation histories.

Whilst WMD clearly has a number of advantages over the use of individually calibrated dates, van de Plassche et al. (2001) highlight a number of limitations that should be considered when applying the technique to salt-marsh
sequences. The results of WMD are influenced by the number, frequency and
distribution of available dates, and the characteristics of the corresponding
portion of the ^{14}C calibration curve. Furthermore, wiggle-matched chronologies are not unique solutions since the manner in which data are manipulated

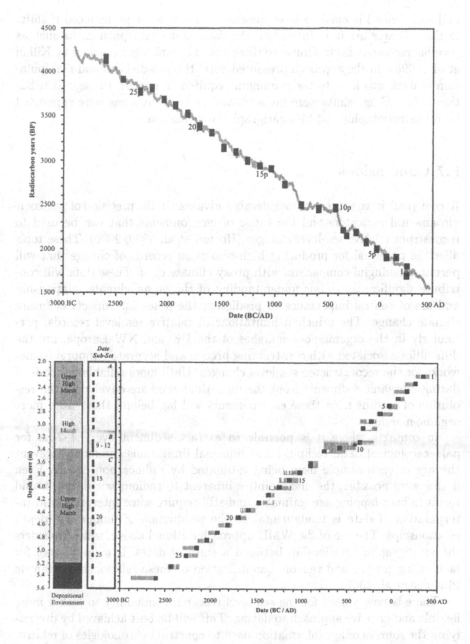

Fig. 9.6. (top) AMS radiocarbon wiggle-matched dates from Pattagansett River Marsh, Connecticut, USA, plotted against the ^{14}C calibration curve (adapted from van de Plassche et al. 2001). (bottom) AMS radiocarbon wiggle-matched dates from Pattagansett River Marsh, Connecticut, USA, presenting the tri-partite sub-division of dates (adapted from van de Plassche et al. 2001). The dark squares show the wiggle-matched dates whilst the pale grey boxes indicate the individually calibrated dates (1σ error range).

and sub-divided is open to interpretation. This is most pronounced if shifts in the ^{14}C ages are introduced into the data to describe phenomena such as possible reservoir effects similar to those noted in ombrogenous bogs by Kilian et al. (1995). In the approach presented here, the sub-division and manipulation of data was kept to the minimum required to explain the age distributions. No ^{14}C age-shifts were invoked, and the sub-divisions were supported by lithostratigraphic and biostratigraphic information.

9.7 Conclusions

Recent years have witnessed considerable advances in the precision of palaeoenvironmental indicators and the range of environments that can be used to reconstruct relative sea-level changes (Horton et al. 1999, 2000). These tools offer the potential for producing high-resolution records of change that will permit meaningful comparison with proxy climate data. These data will contribute significantly to our understanding of the ocean–climate relationship and are of central importance to predicting the potential impacts of future climate change. The principal limitations of relative sea-level records, particularly in the organic-poor marshes of the UK and NW Europe, are the difficulties associated with constructing precise and accurate temporal frameworks for the reconstructed sea-level changes. Until more reliable methods of dating inorganic sediments from the inter-tidal zone are developed, the resolution of records from these environments will lag behind those from more organic marshes.

In contexts where it is possible to extract sediment every 1–2 cm for palaeoecological data, technical and financial limitations commonly prevent the age of each sample from being estimated by radiocarbon dating. Even if this were possible, the uncertainties inherent to radiometric dating would result in over-lapping age estimates and still require some interpretation. Interpolation of data is fundamental to the production of unique age–depth relationships. The use of the WMD approach outlined above, which considers the stratigraphic relationship between a suite of dates, is a useful tool for facilitating precise and reliable quantification of these relationships (van de Plassche et al. 2001).

There is now a need for the sea-level research community to seek a more flexible and creative approach to dating. This will be best achieved by diversifying the sources of age information used to construct chronologies of relative sea-level change. Perhaps the greatest rewards rest with the development of interdisciplinary research projects, such as the fusion with archaeology mentioned above. It may be that human interference and activity in the coastal zone, which has often hampered late Holocene sea-level research, will ultimately provide the solution to one of its outstanding problems.

Acknowledgements

I gratefully acknowledge the meticulous work of O. van de Plassche in the collection and preparation of the AMS radiocarbon dates from Pattagansett River Marsh, and thank A. Speranza for guidance in using *Cal25*. I am also grateful to my former colleagues in the Department of Geography, University of Durham, for the stimulating discussions that have helped shape some of the ideas presented in this chapter. This chapter benefited from the thoughtful comments of Prof. Michael Tooley and Dr. Robert Van de Noort.

References

Aitken, M. J. (1990). *Science-based dating in archaeology*. Longman, London.

Allen, J. R. L. (1987a). Coal dust in the Severn Estuary, southwestern UK. *Marine Pollution Bulletin*, **18**, 169–174.

Allen, J. R. L. (1987b). Toward a quantitative chemostratigraphic model for sediments of late Flandrian age in the Severn Estuary, UK. *Sedimentary Geology*, **53**, 73–100.

Allen, J. R. L. (1988). Modern-period muddy sediments in the Severn Estuary (southwestern UK): a pollutant-based model for dating and correlation. *Sedimentary Geology*, **58**, 1–21.

Allen, J. R. L. (1990a). Constraints on measurement of sea-level movements from salt marsh accretion rates. *Journal of the Geological Society of London*, **147**, 5–7.

Allen, J. R. L. (1990b). The formation of coastal peat marshes under an upward tendency of relative sea-level. *Journal of the Geological Society of London*, **147**, 743–745.

Allen, J. R. L. (1990c). Salt-marsh growth and stratification; a numerical model with special reference to the Severn Estuary, southwest Britain. *Marine Geology*, **95**, 77–96.

Allen, J. R. L. (1991). Salt-marsh accretion and sea-level movement in the inner Severn Estuary, southwest Britain: the archaeological and historical contribution. *Journal of the Geological Society of London*, **148**, 485–494.

Allen, J. R. L. and Rae, J. E. (1986). Time sequence of metal pollution, Severn Estuary, southwestern UK. *Marine Pollution Bulletin*, **17**, 427–431.

Bailiff, I. K. and Tooley, M. J. (2000). Luminescence dating of fine-grain Holocene sediments from a coastal setting. In I. Shennan and J. E. Andrews (eds.), *Holocene land–ocean interaction and environmental change around the western North Sea*, Geological Society Special Publications, London, vol. 166, 55–67.

Chapman, V. J. (1960). *Saltmarshes and salt deserts of the world*. Hill, London.

Clarke, M. L. and Rendell, H. M. (2000). The development of a methodology for luminescence dating of Holocene sediments at the land–ocean interface.

In I. Shennan and J. E. Andrews (eds.), *Holocene land–ocean interaction and environmental change around the western North Sea*, Geological Society Special Publications, London, vol. 166, 69–86.

Cundy, A. B. and Croudace, I. W. (1996). Sediment accretion and recent sea-level rise in the Solent, southern England – inferences from radiometric and geochemical studies. *Estuarine Coastal and Shelf Science*, **43**, 449–467.

Devoy, R. J. N. (1979). Flandrian sea level changes and vegetational history of the lower Thames estuary. *Philosophical Transactions of the Royal Society of London*, **285B**, 355–410.

Edwards, R. J. (2001). Mid to late Holocene relative sea-level change in the Hampshire Basin, UK: new data from Poole Harbour. *Journal of Quaternary Science*, **16**, 221–235.

Fairbanks, R. G. (1989). A 17,000-year glacio-eustatic sea level record: influence of glacial melting rates on the Younger Dryas event and deep ocean circulation. *Nature*, **342**, 637–642.

Fulford, M., Champion, T. and Long, A. J. (1997). *England's coastal heritage*. Coastal Heritage and the Royal Commission on the Historical Monuments of England, London.

Gehrels, W. R. (1999). Middle and late Holocene sea-level changes in eastern Maine reconstructed from foraminiferal saltmarsh stratigraphy and AMS 14C dates on basal peat. *Quaternary Research*, **52**, 350–359.

Gehrels, W. R., Belknap, D. F., Black, S. and Newham, R. M. (2002). Rapid sea-level rise in the Gulf of Maine, USA, since AD 1800. *The Holocene*, **12**, 383–389.

Godwin, H. (1940). Studies of the post-glacial history of the British vegetation. III Fenland Pollen Diagrams. IV Post-glacial changes in relative land- and sea-level in the English Fenland. *Philosophical Transactions of the Royal Society*, **B570**, 239–303.

Godwin, H. (1945). Coastal peat beds of the North sea region, as indices of land- and sea-level changes. *New Phytologist*, **44**, 26–29.

Godwin, H. and Godwin, M. E. (1940). Submerged peat at Southampton — data for the study of post-glacial history V. *New Phytologist*, **39**, 303–307.

Godwin, H., Suggate, R. P. and Willis, E. H. (1958). Radiocarbon dating of the eustatic rise in ocean-level. *Nature*, **181**, 1518–1519.

Gray, A. J. (1992). Saltmarsh plant ecology: zonation and succession revisited. In J. R. L. Allen and K. Pye (eds.), *Saltmarshes: morphodynamics, conservation and engineering significance*, Cambridge University Press, Cambridge, 63–79.

Horton, B. P., Edwards, R. J. and Lloyd, J. M. (1999). A foraminiferal-based transfer function: implications for sea-level studies. *Journal of Foraminiferal Research*, **29**, 117–129.

Horton, B. P., Edwards, R. J. and Lloyd, J. M. (2000). Implications of a microfossil transfer function in Holocene sea-level studies. In I. Shennan and J. E. Andrews (eds.), *Holocene land–ocean interaction and environ-*

mental change around the western North Sea, Geological Society Special Publications, London, vol. 166, 41–54.

Kilian, M. R., van der Plicht, J. and van Geel, B. (1995). Dating raised bogs: new aspects of AMS 14C wiggle matching, a reservoir effect and climate change. *Quaternary Science Reviews*, **14**, 959–966.

Laborel, J., Morhange, C., Lafont, R., Le Champion, J., Laborel-Deguen, F. and Sartoretto, S. (1994). Biological evidence of sea-level rise during the last 4500 years on the rocky coasts of continental southwestern France and Corsica. *Marine Geology*, **120**, 203–233.

Long, A. J. (1992). Coastal responses to changes in sea-level in the East Kent Fens and southeast England, UK over the last 7500 years. *Proceedings of the Geologists' Association*, **103**, 187–199.

Long, A. J. and Roberts, D. H. (2002). A revised chronology for the Fjord Stade moraine in Disko Bugt, west Greenland. *Journal of Quaternary Science*, **17**, 561–579.

Long, A. J., Scaife, R. G. and Edwards, R. J. (1999). Pine pollen in intertidal sediments from Poole Harbour, UK: implications for late-Holocene sediment accretion rates and sea-level rise. *Quaternary International*, **55**, 3–16.

Long, A. J., Scaife, R. G. and Edwards, R. J. (2000). Stratigraphic architecture, relative sea-level, and models of estuary development in southern England: new data from Southampton Water. In K. Pye and J. R. L. Allen (eds.), *Coastal and estuarine environments: sedimentology, geomorphology and geoarchaeology*, Geological Society Special Publications, London, vol. 175, 253–279.

Long, A. J. and Tooley, M. J. (1995). Holocene sea-level and crustal movements in Hampshire and Southeast England, United Kingdom. *Journal of Coastal Research, Special Issue*, **17**, 299–310.

Lowe, J. J. and Walker, M. J. C. (1997). *Reconstructing Quaternary environments*. Prentice Hall, Harlow.

Mastronuzzi, G. and Sansò, P. (2002). Holocene uplift rates and historical rapid sea-level changes at the Gargano promontory, Italy. *Journal of Quaternary Science*, **17**, 593–606.

Mook, W. G. and van de Plassche, O. (1986). Radiocarbon dating. In O. van de Plassche (ed.), *Sea level research: a manual for the collection and interpretation of data*, Geo Books, Norwich, 525–560.

Pethick, J. S. (1980). Salt marsh initiation during the Holocene transgression: the example of the north Norfolk marshes, England. *Journal of Biogeography*, **7**, 1–9.

Pethick, J. S. (1981). Long-term accretion rates on tidal salt marshes. *Journal of Sedimentary Petrology*, **51**, 571–577.

Pirazzoli, P. A. (1986). Marine notches. In O. van de Plassche (ed.), *Sea level research: a manual for the collection and interpretation of data*, Geo Books, Norwich, 361–400.

Pirazzoli, P. A. (1996). *Sea-level changes: the last 20,000 years*. Wiley, Chichester.

Pizzuto, J. E. and Schwendt, A. E. (1997). Mathematical modeling of autocompaction of a Holocene transgressive valley-fill deposit, Wolfe Glade, Delaware. *Geology*, **25**, 57–60.

Pye, K. and Allen, J. R. L. (2000). Past, present and future interactions, management challenges and research needs in coastal and estuarine environments. In K. Pye and J. R. L. Allen (eds.), *Coastal and estuarine environments: sedimentology, geomorphology and geoarchaeology*, Geological Society Special Publications, London, vol. 175, 1–4.

Sawai, Y., Nasu, H. and Yasuda, Y. (2002). Fluctuations in relative sea-level during the past 3000 yr in the Onnetoh estuary, Hokkaido, northern Japan. *Journal of Coastal Research*, **17**, 607–622.

Scott, D. B. and Medioli, F. S. (1980). *Quantitative studies of marsh foraminiferal distributions in Nova Scotia: implications for sea level studies*. Cushman Foundation for Foraminiferal Research, DeKalb, Illinois.

Shahidul Islam, M. and Tooley, M. J. (1999). Coastal and sea-level changes during the Holocene in Bangladesh. *Quaternary International*, **55**, 61–75.

Shennan, I. (1982). Interpretation of the Flandrian sea-level data from the Fenland, England. *Proceedings of the Geologists Association*, **93**, 53–63.

Shennan, I. (1986). Flandrian sea-level changes in the Fenland II: tendencies of sea-level movement, altitudinal changes, and local and regional factors. *Journal of Quaternary Science*, **1**, 155–179.

Shennan, I. (1994). Coastal evolution. In M. Waller (ed.), *The Fenland projects, number 9: Flandrian environmental change in the Fenland*, East Anglian Archaeology, Cambridge, 47–84.

Shennan, I. and Horton, B. P. (2002). Holocene land- and sea-level changes in Great Britain. *Journal of Quaternary Research*, **17**, 511–526.

Shennan, I., Innes, J. B., Long, A. J. and Zong, Y. (1995). Late Devensian and Holocene relative sea-level changes in northwestern Scotland: new data to test existing models. *Quaternary International*, **26**, 97–123.

Shennan, I., Lambeck, K., Flather, R., Horton, B., Mcarthur, J., Innes, J., Lloyd, J., Rutherford, M. and Wingfield, R. (2000). Modelling western North Sea palaeogeographies and tidal changes during the Holocene. In *Holocene land–ocean interaction and environmental change around the North Sea*, Geological Society Special Publications, London, vol. 166, 299–319.

Shennan, I., Tooley, M. J., Davis, M. J. and Haggart, B. A. (1983). Analysis and interpretation of Holocene sea-level data. *Nature*, **302**, 404–406.

Stuiver, M., Reimer, P. J., Bard, E., Beck, J. W., Burr, G. S., Hughen, K. A., Kromer, B., McCormac, F. G., Plicht, J. V. D. and Spurk, M. (1998). IntCal98 radiocarbon age calibration, 24,000-0 cal BP. *Radiocarbon*, **40**, 1041–1083.

Sutherland, D. G. (1982). Dating and associated methodological problems in the study of Quaternary sea-level changes. In A. F. Harding (ed.), *Climate change in later prehistory*, Edinburgh University Press, Edinburgh, 165–197.

Switsur, R. (1994). Methods of reconstruction: IV. radiocarbon dating. In M. Waller (ed.), *The Fenland projects, number 9: Flandrian environmental change in the Fenland*, East Anglian Archaeology, Cambridge, 27–34.

Turney, C. S. M., Coope, G. R., Harkness, D. D., Lowe, J. J. and Walker, M. J. C. (2000). Implications for the dating of Wisconsinan (Weichselian) late-glacial events of systematic radiocarbon age differences between terrestrial plant macrofossils from a site in SW Ireland. *Quaternary Research*, **53**, 114–121.

van de Plassche, O. (1982). *Sea-level and water-level movements in the Netherlands during the Holocene*. Geological Survey of the Netherlands, Amsterdam.

van de Plassche, O. (1986). *Sea-level research: a manual for the collection and evaluation of data*. Geo Books, Norwich.

van de Plassche, O. (2000). North Atlantic climate–ocean variations and sea level in Long Island Sound, Connecticut, since 500 cal yr AD. *Quaternary Research*, **53**, 89–97.

van de Plassche, O., Edwards, R. J., van der Borg, K. and De Jong, A. F. M. (2001). 14C wiggle-match dating in high-resolution sea-level research. *Radiocarbon*, **14**, 391–402.

van der Plicht, J. (1993). The Groningen radiocarbon calibration program. *Radiocarbon*, **35**, 231–237.

van Geel, B. and Mook, W. G. (1989). High-resolution 14C dating of organic deposits using natural atmospheric 14C variations. *Radiocarbon*, **31**, 151–156.

Varekamp, J. C., Thomas, E. and van de Plassche, O. (1992). Relative sea-level rise and climate change over the last 1500 years. *Terra Nova*, **4**, 293–304.

Waller, M. (1994). Coastal evolution. In M. Waller (ed.), *The Fenland projects, number 9: Flandrian environmental change in the Fenland*, East Anglian Archaeology, Cambridge, 47–84.

10

A Framework for Analysing Fossil Record Data

Robert E. Weiss, Sanjib Basu, and Charles R. Marshall

Summary. This chapter focuses on a new approach to building chronologies on the basis of data from stratigraphic sequences of fossil plants and/or animals sealed within geological deposits. In order to understand the speed and timing of changes within the fossil record, geologists have for many years made systematic studies of the nature and number of taxa present in geological sequences. However, due to the incompleteness of the fossil record, a particular taxon may not be observed even when it is extant at a particular sampling point. Sampling intensity can vary across sampling points by orders of magnitude, and, depending on appearances (originations) and extinctions, different taxa compete to become part of the sample. This chapter offers a Bayesian statistical framework for interpreting data of this type. Abundance and depth (or stratigraphic position) data are combined to estimate the times of appearances and disappearances of taxa in the presence of prior information, including an estimated longevity of each taxon and the probability that it will be observed if extant.

10.1 Introduction

Fossils are commonly collected by paleontologists to help answer questions about the history of life. When detailed chronological data are desired, for example when assessing the rapidity or taxonomic scope of a mass extinction, samples may be collected systematically from successively younger stratigraphic layers (horizons) at specific localities or sites (e.g. Marshall and Ward 1996). These layers may be exposed in rock outcrops such as road cuts or quarry walls, etc., or may be recovered from drill cores. Through systematic collection, a temporal sequence of data points is accumulated, each recovered fossil lying some stratigraphic distance above (or below) some reference stratigraphic horizon. From these data, the patterns of local appearance and extinction of the taxa (usually species or genera) at the collected locality may then be inferred.

However, stratigraphic (chronological) gaps are found in the ranges of the taxa recovered from even systematically collected stratigraphic sections. This incompleteness of the fossil record often seriously hampers our ability to read the chronology of life directly from the fossil record. For example, it has been shown that sudden extinctions typically appear gradual in the fossil record as a consequence of the incompleteness of the fossil record (the Signor–Lipps effect) (Signor and Lipps 1982; Alvarez 1984; Raup 1989; Springer 1990; Marshall 1995, 1998).

Here we are interested in statistical analysis of the stratigraphic distributions of fossils recovered from systematically collected stratigraphic sections. Our goal is to tease out, to the best of our ability, the true chronological patterns of appearances and extinctions from those patterns that may result from the incompleteness of the fossil record or from the sampling regime used.

10.1.1 What We Hope to Learn from the Fossil Record

Often detailed paleontological data are collected from localities representing the same time interval from geographically dispersed localities. Typically, we would like to infer

1. the times of appearances and extinction and measures of uncertainty for these times for each taxon at each location;
2. probabilities of co-appearance and co-extinction of taxa at each location;
3. probabilities of co-appearance and co-extinction of the same taxon between localities; and
4. whether there were changes in the relative abundances of taxa at particular times, or between different geographic areas.

10.1.2 Sampling Regimes

Detailed stratigraphic data may be collected from stratigraphic sections in one of several ways as follows.

Continuous Sampling

Fossils are sampled by continuous examination of exposed rock layers. The stratigraphic position and taxonomic identity of each fossil found is recorded. This strategy is often used for macro-fossils, such as clams, ammonites or trilobites, recovered from exposed outcrops (e.g. Marshall and Ward 1996).

Point Sampling

Rock samples are collected at a series of stratigraphic points up an outcrop or drill core. A count is made of the observed abundance of the different taxa at each sampled point (generally each sample has a finite stratigraphic thickness, but in point sampling this thickness is small when compared with the stratigraphic distance between sample points). This strategy is common when sampling micro-fossils from drill cores or outcrops (e.g. see unpublished foraminiferal data of Vaziri 1997, used by Weiss and Marshall 1999).

Binned Sampling

Fixed volumes of rock, usually of standard stratigraphic thickness, are collected, and a count is made of the different taxa and their abundances in each sampled bin. Generally, the positions of the fossils within each sample is not recorded. In places where the time interval of interest is stratigraphically thin (i.e. represented by metres, rather than by tens or hundreds of metres of rock), no stratigraphic gaps are left between successive bins (e.g. Thomas 1993, 1995, see below); that is, the samples are stratigraphically contiguous. However, for thick sections, there are often stratigraphic gaps left between the sampled bins (e.g. Jin et al. 2000). Binned sampling is usually employed for macro-fossils recovered from outcrops.

10.1.3 Trilobite Data

The fossil record indicates that the history of animal life has been punctuated by several extremely large mass extinctions (Sepkoski 1997) as well as by a series of smaller "minor" mass extinctions. The earliest of these minor mass extinctions reported in North American rocks are of trilobite fauna during the Cambrian Period (from 543 to 490 million years ago) (Palmer 1965, 1984, 1998; Stitt 1971). There are several such events that occurred in the Cambrian, each largely extinguishing the standing crop of trilobite genera alive at the time. The trilobite genera found between successive extinction events are collectively known as a biomere. The biomere boundary is the ending of one biomere and the beginning of the next.

Thomas (1993) used binned (contiguous) sampling across the boundary between the Marjumiid and Pterocephaliid (Steptoean) biomeres (in the Late Cambrian) to assess the rapidity of the Marjumiid biomere extinction event. He documented in detail 13 localities across the Western United States, in Montana, Wyoming, Utah, Texas and Nevada. In almost all cases he collected a fixed volume of rock, ten centimetres (cm) thick for each sample. The most notable exception is a one cm thick bin typically sampled immediately below the biomere boundary. Thirty-four genera of trilobite were collected over the 13 sections, although most are rare both regionally and locally.

In Tables 10.1 and 10.2 we show data from two of the seven sites that serve as exemplars for this chapter. These are the Beartooth Butte and Cathedral Cliffs sites, both from Wyoming (Thomas 1993). Initial inspection of the data from the 13 sites suggested that seven had similar patterns of appearance and disappearance. These seven sites are Beartooth Butte, WY; Cathedral Cliffs, WY; Fox Creek, WY; Highland Range, NV; Little Horse Canyon, UT; Morgan Creek, TX; and North Fork Grove Creek, MT.

216   Robert E. Weiss, Sanjib Basu, and Charles R. Marshall

Table 10.1. Data from Beartooth Butte, WY. *Height* is the height in metres from a baseline. *Total* is the total number of specimens collected from that horizon. Remaining columns are counts of that particular taxon. Blanks indicate zero counts. *Aph.* = *Aphelaspis*, *Tri.* = *Tricrepicephalus*, *Con.* = *Coosina*, *Gla.* = *Glaphyraspis*, *Cre.* = *Crepicephalus*, *Che.* = *Cheilocephalus*, *Oth.* = sum of counts of 28 additional taxa. The Marjumiid–Pterocephaliid biomere boundary is located as indicated by the horizontal line at 2.60 m above the baseline above the 6th and below the 7th horizon.

Horizon	Height	Total	*Aph.*	*Tri.*	*Con.*	*Gla.*	*Cre.*	*Che.*	*Oth.*
18	5.50	123	123						
17	5.20	70	20				50		
16	5.00	484	471				13		
15	4.80	40	40						
14	4.60	20	18				1	1	
13	4.00	21	21						
12	3.70	11	10				1		
11	3.50	11	9				1	1	
10	3.30	229	224					5	
9	3.10	123	116				1	6	
8	2.70	56					56		
7	2.60	47					47		
6	2.59	41		31	10				
5	2.20	3		1	1				1
4	1.50	15		5	10				
3	1.30	4		1	3				
2	1.20	32		8	14			8	2
1	0.10	1							1

Each row in Tables 10.1 and 10.2 represents a sampling bin, and is here called a *horizon*; 18 horizons were sampled at Beartooth Butte, and 19 at Cathedral Cliffs. The second column gives the height t_j of each horizon above an arbitrary baseline in metres. Height is a proxy for time; we will generally use the imagery of time in the sequel. Excepting right below the biomere boundary, each sampling bin was 10 cm thick. At Beartooth Butte the biomere boundary is above horizon 6 and below 7 at $t_j = 2.60$. For Cathedral Cliffs, the biomere boundary is above horizon 9 and below 10 at $t_j = 6.70$. The third column contains the total number of trilobite specimens (fossils) found at each horizon. The remaining columns show the number of specimens from each of the six most abundant taxa; the last column gives the total number of fossils from other, rare, taxa. At Beartooth Butte these represent just four genera each known from one fossil each, while at Cathedral Cliffs these represent four genera known from a total of six specimens. Of these rare genera, only one is held in common between the two localities. Blanks in the tables indicate that zero specimens were recovered.

Table 10.2. Data from Cathedral Cliffs, WY. Format and definitions are the same as for Table 10.1. The Marjumiid–Pterocephaliid biomere boundary is located as indicated by the horizontal line at 6.70 m above baseline above the 9th and below the 10th horizon.

Horizon	Height	Total	Aph.	Tri.	Con.	Gla.	Cre.	Che.	Oth.
19	8.80	97	92			5			
18	8.50	141	132			9			
17	8.40	27	24			3			
16	8.30	81	76			5			
15	8.20	483	460			23			
14	8.00	88	64			24			
13	7.30	430	387			28		15	
12	7.00	5				5		1	
11	6.80	23				23		1	
10	6.70	52				52			
9	6.69	20		15	5				
8	6.60	32		9	23				
7	6.20	8		3	5				
6	6.10	34		12	21				1
5	6.00	16		6	9				1
4	4.80	10		6	4				
3	4.00	62		17	45				
2	3.50	69			69				
1	0.10	68		8	44	12			4

In our analyses we exclude the rare taxa, and thus for our calculations we use only the total number of fossils recovered for the six common genera. We treat the samples as if they were collected from the exact point given in the first column; that is, we treat the binned sampling as if it were point sampling; at this time we have not extended our models to bin sampling. The samples at the point 2.59 metres at Beartooth Butte and 6.69 metres at Cathedral Cliffs come from bins of 1 cm height that immediately underlay the Marjumiid–Pterocephaliid biomere boundary. For all 10 cm thick sampling intervals the sampling intensity was equal; that is, a fixed volume of rock was disaggregated and all the fossils recovered. However, some horizons are more fossiliferous than others as indicated by the wide range of number of specimens (1 to 484) collected in each bin at Beartooth Butte.

For these data interesting questions include the following. 1) Was there a simultaneous extinction at or near the biomere boundary, and if so, which taxa were involved? 2) If the extinction was not simultaneous, over what stratigraphic interval did the genera really become extinct? 3) Further, were all genera already extant at the base of the sections, or did some appear/immigrate some time after the base of the section? 4) Did the same taxa appear and become extinct at the same times in multiple sections? 5) Finally, what were the significant changes in relative abundance through each section?

218   Robert E. Weiss, Sanjib Basu, and Charles R. Marshall

10.1.4 Previous Statistical Methods of Compensating for the Incompleteness of the Fossil Record

There are relatively few formal statistical models for paleontological data of this type. Following the pioneering work of Shaw (1964) and Paul (1982), Strauss and Sadler (1989) provided the first statistical model suitable for macro-fossil finds (see Marshall 1998, for a review). Other contributions include Marshall (1994, 1997), Solow (1996a,b), Solow and Smith (1997, 2000), Wang and Marshall (in preparation), and Weiss and Marshall (1999). Strauss and Sadler (1989), Solow (1996a) and Weiss and Marshall (1999) employ Bayesian approaches with Strauss and Sadler (1989) also giving classical approaches. Strauss and Sadler (1989) implicitly assume a constant collecting intensity from minus infinity to plus infinity along the stratigraphic column. Their analysis is for continuous sampling and incorporates only the locations of the first and last finds, and a count of the number of interior locations where fossils are found.

Weiss and Marshall (1999), hereafter WM99, began with a model for the extinction of a single taxon assumed extant at the base of the section. Their data were Bernoulli presence/absence observations at times t_j above bottom. At some point, the taxon is no longer observed in the fossil record. This could be due to false negatives, a lack of further sampling, or due to the occurrence of local extinction. The model of WM99 accounts for the possibility of false negatives in developing a posterior distribution for the extinction time D_i (S in WM99) of a taxon. WM99 accounts for the locations where data have been sampled and where it cannot be sampled. It is suitable for analysis of microfossils where we collect and identify many samples at a single stratigraphic location. Marshall (1997) provides a method based on a generalization of Strauss and Sadler's classical approach that can accommodate sampling information into the estimate of the true time of appearance or extinction.

In this chapter we propose a framework for analysing typical fossil record data. There are numerous variants in the way basic paleontological data are collected and similarly there are numerous questions of interest in real situations. Our model and its variations can handle many of them and we illustrate various inferences using our model.

Key features of the data set are n_j, the number of fossils identified at horizon t_j. The observations y_{ij} are the actual specimen counts of the I taxa at horizon t_j. We study multiple taxa in a single analysis. Our interests lie in identifying the appearance and extinction times of taxa, measures of uncertainty for those times and measures of joint extinction or appearance.

Our models are appropriate for both micro-fossils or macro-fossils sampled at discrete locations (point sampling) up the fossiliferous outcrop or core. We introduce inferences about simultaneous extinction or appearance of two or more taxa at a single site. Our model can handle simultaneous extinction of the same or different taxa at different sites as well, as long as an independent temporal reference frame can be provided.

The next section presents our basic continuous time model. The unknown parameters in our models are (i) the appearance and extinction times for each taxon at each site and (ii) the relative abundances of the taxa. Relative abundance of taxa at a single horizon determine the probability that a single specimen drawn belongs to each taxon. We describe our model and priors in mathematical detail, and then discuss a modification where we model the appearance and extinction times as falling in discrete chunks or bins rather than fitting appearance and extinction times as continuous parameters. We apply our methods to the Thomas data in Section 10.3 and we end with discussion in Section 10.4.

10.2 Models and Priors

For each taxon at a particular site, our model has three parameters. Two parameters are the appearance and extinction times for the taxon. The third parameter is the relative abundance of the taxon. Taxa that are relatively abundant will have a large value of this parameter, while relatively rare taxa will have a small value. Our model says that each specimen identified at a particular horizon has a probability of being identified as a particular taxon. The probability is non-zero if the taxon is extant at that horizon, and the probability depends on the relative population size as compared with the other extant taxa.

We next describe our model mathematically, then we describe three prior assumptions about the abundances across sites.

10.2.1 The Model

Our model analyses taxa counts y_{ijk} for taxa i with i running from 1 to I at horizon t_j with j running from 1 to J_k with $t_{jk} < t_{(j+1)k}$ within site k. There are K sites, I taxa and J_k time points where data have been observed at site k. We may eliminate rare or irrelevant taxa from consideration, reducing n_j and the taxon count I. In the current model and data set we have the same taxa at all sites but this is not strictly necessary.

We wish to estimate the extinction time D_{ik}, and the appearance time B_{ik} with $B_{ik} < D_{ik}$ for each taxon at each site. Define the existence function $S_{ik}(t)$

$$S_{ik}(t) = \mathbf{1}\{B_{ik} \leq t < D_{ik}\},$$

which is a 0–1 indicator function and let $S_{ijk} = S_{ik}(t_{jk})$ be the indicator function evaluated at time t_{jk} and the vector of indicators $S_{jk} = (S_{1jk}, \ldots, S_{Ijk})^t$ distinguished from the function by the arguments and the names of the subscripts. The vector of taxa counts for time j, site k is $y_{jk} = (y_{1jk}, \ldots, y_{Ijk})^t$ which will be multinomial with parameters n_{jk} and $\pi_{jk} = (\pi_{1jk}, \ldots, \pi_{Ijk})^t$, where n_{jk} is the known total specimen count at time t_{jk} and site k and the probabilities π_{ijk} are yet to be defined. For $t_{jk} < B_{ik}$ or $t_j \geq D_{ik}$, we have $y_{ijk} = 0$ with probability one.

To define π_{ijk}, we introduce a_{ijk} which may be thought of as the relative abundance or relative population size of taxon i at time t_j and site k presuming that the taxon is extant. We separate existence from abundance; $a_{ijk} > 0$ at all times. We define the probability that a single specimen is taxa i at time j, site k to be

$$\pi_{ijk} = \frac{a_{ijk}S_{ijk}}{\sum_{m:\text{ all extant taxa }} a_{mjk}}$$
$$= \frac{a_{ijk}S_{ijk}}{a_{jk}^t S_{jk}}, \tag{10.1}$$

where $a_{jk} = (a_{1jk}, \ldots, a_{Ijk})^t$. For inferential purposes, important quantities are relative sizes (hence relative abundance) of a_{ijk}s, that is, ratios of a_{ijk}s which are potentially determined by the data. For example, if we multiply all a_{ijk} by a constant, the model will not change, consequently we fix the sum over i of the a_{ijk} to be $1 = \sum_i a_{ijk}$. The ratio $a_{ijk} / \sum_i a_{ijk}$ can be thought of as the per-specimen probability of observing taxon i at a time when all taxa are extant simultaneously. For brevity, we call the a parameters abundances.

To get a model which is (a) fittable, (b) not too complex and (c) hopefully not too inappropriate for our data, we will assume that $a_{ijk} \equiv a_{ik}$ with site-specific abundances $a_k = (a_{1k}, \ldots, a_{Ik})$ constant across different times t_j within site k.

Our model sets an exponential prior on the appearance and extinction times of taxa, a Dirichlet prior on a_k and a multinomial distribution for the vector of taxon counts y_{jk}:

$$a_{jk} \equiv a_k$$
$$a_k \sim \text{Dirichlet}(A_{00}, N_{00}) \tag{10.2}$$
$$y_{jk} \sim \text{Multinomial}(\pi_{jk}, n_{jk}) \tag{10.3}$$
$$t_{J_k k} - B_{ik} \sim \text{Exponential}(\mu_B) \tag{10.4}$$
$$D_{ik} - t_{1k} \sim \text{Exponential}(\mu_D), \tag{10.5}$$

where $\mu_B = \mu_D = \mu$ are known prior parameters on the lifespans of taxa. It might be useful to take $\mu_B \neq \mu_D$ in some data sets, and definitely it might be useful to change them across sites and taxa, although we do not. We have been able to elicit estimates from experts on "typical" lifespan for taxa. We then halve those values and set μ equal to the resultant.

To fully specify our model, we need the definition of π_{ijk} from Equation 10.1, we need to specify A_{00} and N_{00} and we need Equations 10.2–10.5. In all models, N_{00} is assumed known.

10.2.2 Three Priors

We consider three different prior models. These models define the relationships across sites of the within-site parameters. Because we have only a single horizon that is identified across sites, we do not tie together appearance and extinction times across sites; knowledge of the appearance or extinction times at one site do not inform the appearance or extinction times at another site. Thus, as far as appearance and extinction times are concerned, the sites are treated independently.

We consider several models for the relationship of abundances across sites. In model 1, the independent model, we assume that there is no information at one site that reflects on abundance at other sites. In the opposite model, model 2, we assume that abundances are equal at all seven sites. Finally, in model 3, we assume *exchangeability* of abundances across sites. In this model, we assume that there is an overall unknown average abundance across sites. Abundances within sites are close to the average, but are not the same across sites.

Model 1. The independent a_k model. In this model, sites are independent and estimation proceeds within each site separately.

$$a_k \sim \text{Dirichlet}(A_{00}, N_{00})$$

where A_{00} is known.

Model 2. The equal a model.

$$a_k \equiv a$$

$$a \sim \text{Dirichlet}(A_{00}, N_{00})$$

with A_{00} and N_{00} known. The a vector is shared across all times and all sites. This may be appropriate for physically close sites, but may not be appropriate for our data where sites are spread over fairly large distances.

Model 3. The exchangeable a_k model.

$$a_k \sim \text{Dirichlet}(A_{00}, N_{00})$$

$$A_{00} \sim \text{Dirichlet}(A_{11}, N_{11}),$$

with A_{11} and N_{11} known. This is an exchangeable prior which allows information about abundances from one site to assist in the estimation of abundance at other sites. The double Dirichlet prior is not conjugate; other priors certainly could be entertained.

10.2.3 Discrete Time

Appearances B_{ik} and extinctions D_{ik} are continuous variables, however, the likelihoods for B_{ik} or D_{ik} are flat in between sampling locations t_{jk} and $t_{(j+1)k}$, where we define $t_{0k} = -\infty$ and $t_{(J_k+1)k} = +\infty$. Posteriors are not flat on intervals $(t_{jk}, t_{(j+1)k})$ solely due to the prior. There is no particular advantage to continuing to treat B_{ik} and D_{ik} as continuous variables; instead we treat the block of time $(t_{jk}, t_{(j+1)k})$ as the primary inference unit. One would report $P(B_{ik} \in (t_{(j-1)k}, t_{jk}]|Y)$ for j running from 1 to the time that taxa i was first observed at site k, and $P(D_{ik} \in [t_{jk}, t_{(j+1)k})|Y)$ for j running from the time of the last observation of taxa i at site k up to J.

In fact, one can make the bin $(t_{(j-1)k}, t_{jk}]$ of initial appearance and the bin $[t_{jk}, t_{(j+1)k})$ where extinction occurs be the parameter of interest rather than the continuous times B_{ik} and D_{ik} respectively. We call these bin parameters α_{ik} and ω_{ik} for the appearance bin and extinction bin respectively, taking values in $1, 2, \ldots, j_{ik,\text{first}}$ and $j_{ik,\text{last}}, \ldots, J_k$ respectively where $j_{ik,\text{first}}$ is the index value corresponding to the first t_{jk} where taxon i was seen in site k, similarly, $j_{ik,\text{last}}$ is the last index value of the last time taxon i was seen in site k. If $\alpha_{ik} = j$, then $t_{jk} < B_{ik} < t_{(j+1)k}$, and $\omega_{ik} = j$ implies $t_{jk} \leq D_{ik} < t_{(j+1)k}$.

We model α_{ik} and ω_{ik} as multinomially distributed across the set of possible bins, with prior probabilities proportional to the mass given to them by the exponential prior distributions for the continuous time priors for B_{ik} and D_{ik}. We take the a_k and the α_{ik}s and ω_{ik}s to be our unknown parameters.

Computational issues are discussed in Appendix A. Appendix B discusses a specification issue which is of consequence should we wish to do formal Bayesian model selection or testing with these models.

10.3 Example: the Thomas Trilobite Data

For our prior, we elicited from an expert (R. Thomas) a typical taxon survival time of 10 metres and we therefore set the prior mean expected times to appearance and extinction equal to 5, i.e. the means μ_B and μ_D equal 5 in the priors for the appearances and extinctions. This estimated survival time actually varies by a few metres from site to site. However, as shown in WM99, the posteriors on the survival times are not overly sensitive to small changes in the priors on the survival times and we kept the prior means the same for all sites.

We do not have any prior information to suggest that one taxon is more prevalent than another, nor do we have much information about the relative abundances of taxa. This suggests a flat prior over the abundance parameters a. Mathematically, therefore, for all our Dirichlet(A, N) priors with A and N known, we set A proportional to the vector of ones, and we set $N = 6$.

Our MCMC computational method produces many samples of parameter values and we computed the log-likelihood, a measure of model fit, at those values. Based on the log-likelihood, the data do not support model 2 with its equal abundances across sites as compared with the independent or exchangeable abundance models where abundances differ across sites. The independent

Table 10.3. Summary of posteriors of the abundances, a, from the equal a model (i.e. model 2). Posterior means, standard deviations, 2.5 and 97.5 percent points. The last two columns give a 95% confidence interval for the taxon abundances. See Table 10.1 for genus abbreviations.

Taxon	mean	sd	2.5%	97.5%
Aph.	0.570	0.024	0.522	0.616
Tri.	0.136	0.010	0.116	0.156
Con.	0.094	0.007	0.080	0.108
Gla.	0.042	0.002	0.037	0.046
Cre.	0.121	0.010	0.102	0.142
Che.	0.038	0.003	0.032	0.043

Table 10.4. Posterior means of abundances, a_{ik}, from the independent a model (i.e. model 1) for each site. See Table 10.1 for genus abbreviations.

Site	*Aph.*	*Tri.*	*Con.*	*Gla.*	*Cre.*	*Che.*
Beartooth Butte	0.530	0.167	0.138	0.038	0.108	0.018
Cathedral Cliffs	0.901	0.00029	0.00087	0.075	0.0002	0.023
Fox Creek	0.940	0.0033	0.0024	0.025	0.016	0.014
Highland Range	0.402	0.167	0.0041	0.071	0.328	0.029
Little Horse	0.024	0.401	0.173	0.0071	0.382	0.0066
Morgan Creek	0.605	0.0058	0.0079	0.045	0.0016	0.335
North Fork	0.232	0.496	0.255	0.0060	0.0048	0.0062

abundances model (model 1) is more supported than the exchangeable abundances model (model 3), but we expect that finding to be sensitive to details of the prior and choice of specific sites in the analysis. Care needs to be exercised in the selection of sites in the exchangeable model (model 3); sites must really be quite similar for this model to be an improvement over the independent model (model 1); closer inspection of our sites suggested that they were rather more different than realized on initial inspection.

Table 10.3 presents summaries of the posterior of the abundances a_i for the equal a model. Ratios of a values in the table are estimates of the odds of drawing one taxon versus another when sampling a single specimen; assuming of course that both are extent at the time where we sample. When we correlated abundances a_i with appearance or extinction, lower abundance corresponded to greater uncertainty in the time of appearance or extinction. The less likely we are to observe a taxon, the less certain we are about its appearance or extinction time. Table 10.4 gives estimated posterior means for the abundances for the independent a model.

Table 10.5 reports our inferences about the appearance and extinction bins for the independent model for sites Beartooth Butte and Cathedral Cliffs. If a taxon is not listed in the first part of the table, then the probability is 1.0 that it first appeared in the bin immediately preceding its first appear-

Table 10.5. Summary of the posteriors of the appearances and extinction times for Beartooth Butte and Cathedral Cliffs for the independent a model. Table entries present all non-trivial (i.e. not zero or one) probabilities of appearances in the bins before first actual observation of a specimen in the fossil record, rounded to nearest 0.01. The first probability is for actual appearance in the bin immediately preceding the first appearance of the fossil in the fossil record. The second probability is the probability of appearance in the second bin before the first appearance, and so on. Similar results are given for extinctions; the first probability is the probability that the taxon went extinct in the first bin after the taxon's last appearance in our sample, the second probability is the probability that it actually went extinct in the second bin, and so on. See Table 10.1 for genus abbreviations.

Beartooth appearances	
Gla.	0.60 0.31 0.08 0.01
Beartooth extinctions	
Tri.	0.55 0.24 0.18 0.02 0.01
Con.	0.55 0.23 0.19 0.02 0.01
Cre.	0.87 0.13 0.01
Che.	0.94 0.06
Cathedral extinctions	
Tri.	0.22 0.44 0.23 0.07 0.04
Con.	0.26 0.41 0.22 0.07 0.03 0.01
Che.	0.99 0.01

ance in the fossil record. For example, both *Aphelaspis* and *Cheilocephalus* appeared between horizons 8 and 9 with probability 1.0 in this model. Only *Glaphyraspis* at Beartooth Butte had uncertainty in its appearance time, with 0.6 probability of originating in the 1 cm bin between horizons 6 and 7, and 0.31 probability in the 39 cm bin between horizons 5 and 6. Most of the probability is in the bin immediately before the biomere boundary.

Similarly, if a taxon is not listed in the second part of Table 10.5, then the probability is 1.0 that the taxon became extinct in the first bin after its last observation in the fossil record. For example, *Crepicephalus* at Cathedral Cliffs became extinct between horizon 1 and 2. Several taxa have uncertainty in the exact time of extinction at each site; the first probability listed is the probability that the taxon became extinct in the first bin after it last appeared in the fossil record, the second probability is for the second bin and so on. For example, at Beartooth Butte both *Tricrepicephalus* and *Coosina* have probability 0.55 of having become extinct in the 1 cm bin between horizons 6 and 7 and probabilities 0.24 and 0.23 respectively for becoming extinct in the bin immediately above horizon 7. The exact time of extinction seemed more uncertain than for appearances in all models and the independent a_k model generally has more posterior uncertainty in appearance and extinction times as compared with the equal abundances or exchangeable abundances models.

Table 10.6. Probability of extinction for *Tricrepicephalus* (*Tri.*) and *Coosina* (*Con.*) and appearance of *Glaphyraspis* (*Gla*) at the biomere boundary. Results are presented for all three models and all seven sites. E means the taxon was extant across the boundary (so it could not have originated or become extinct at the boundary). Model 1 has independent (I) abundances across sites; model 2 has equal (Eq) abundances and model 3 has exchangeable (Ex) abundances.

Site / Model	Tri. 1(I)	2(Eq)	3(Ex)	Con. 1(I)	2(Eq)	3(Ex)	Gla. 1(I)	2(Eq)	3(Ex)
Beartooth Butte	0.55	1.00	1.00	0.55	1.00	1.00	0.60	0.98	0.07
Cathedral Cliffs	0.22	1.00	1.00	0.26	1.00	1.00	1.00	0.91	0.18
Fox Creek		E		0.62	0.58	0.60	0.94	0.58	0.98
Highland Range	0.02	0.02	0.03	0.00	0.00	0.00	0.23	0.30	0.30
Little Horse	1.00	1.00	1.00	1.00	1.00	1.00	0.03	0.15	0.02
Morgan Creek		E			E		0.89	0.33	0.99
North Fork	1.00	1.00	1.00	1.00	1.00	1.00		E	

Table 10.6 presents for all three models and all seven sites the probability that *Tricrepicephalus* became extinct, that *Coosina* became extinct and that *Glaphyraspis* first appeared at the biomere boundary. For Beartooth Butte, this is the estimated probability that *Tricrepicephalus* and *Coosina* became extinct between horizons 6 and 7, and the probability that *Glaphyraspis* first appeared between horizons 6 and 7. These probabilities at Beartooth are high for all three models. We see greater uncertainty generally for the independence model – this should not be surprising, as it has the least amount of sharing of information across sites. The equal and exchangeable models give generally similar results with occasional exceptions. For *Tricrepicephalus*, we see that for four sites, Beartooth Butte, Cathedral Cliffs, Little Horse and North Fork, we appear to have extinction at the biomere boundary. *Tricrepicephalus* survived across the boundary for two sites, Fox Creek and Morgan Creek, and at one site, Highland Range, *Tricrepicephalus* probably did not go extinct at the biomere boundary where (data not shown) it probably went extinct before the boundary.

Table 10.7 reports, for all three models and all seven sites, the posterior probability that *Tricrepicephalus* and *Coosina* became extinct in the same bin. We see that five sites appear to have high probability of co-extinctions for *Coosina* and *Tricrepicephalus*, with exact probabilities depending on the prior for two sites. Two sites, Fox Creek and Highland Range, are unlikely to have coincident extinction times. These models also allow us to calculate the probabilities that a taxon became extinct at the biomere boundary simultaneously at several sites, but we do not illustrate that here.

Table 10.7. Probability that *Tricrepicephalus* and *Coosina* became extinct in the same bin for all three models, independent *a* model 1 (Ind), equal *a* model 2 (Equal), and exchangeable *a* model 3 (Exch).

Site	Ind	Equal	Exch
Beartooth Butte	0.96	1.0	1.0
Cathedral Cliffs	1.0	1.0	1.0
Fox Creek	0.01	0.02	0.0
Highland Range	0.02	0.01	0.0
Little Horse	0.45	1.0	1.0
Morgan Creek	0.63	0.52	0.998
North Fork	1.0	1.0	1.0

10.4 Discussion

The assumption of constant relative abundance over time is not appropriate for all data sets. We think it will give reasonable results in many circumstances, including in the model and data of WM99. This assumption has been surprisingly successful and useful, and we recommend it as the appropriate entry point for analysis of new data. We leave the extension of our models to time varying a_k to a (planned) future paper. Both continuous change, random effects, oscillatory and generally constant a_k with occasional jumps may all be reasonable.

In our data, we have a clearly identified horizon across sites at one t_{jk} only, the Marjumiid–Pterocephaliid biomere boundary. The boundary was the time of simultaneous extinction of *Tricrepicephalus* and *Coosina* and appearance of *Glaphyraspis* between horizons 6 and 7 at Beartooth Butte (Table 10.1) and is between horizons 9 and 10 at Cathedral Cliffs (Table 10.2). In other data sets, it may be possible to link most or all horizons t_{jk} across sites, thus potentially allowing us to have t_j not only nested within sites but identifiable across sites, allowing for possible communal estimation of single appearance and extinction times across sites, or for testing whether such hypotheses hold.

Our results bear out the primary conclusions drawn by Thomas (1993) based on his paleontologic judgement of the combined fossil records from the sections he studied. In particular, our analyses confirm the general pattern of co-extinction of *Coosina* and *Tricrepicephalus*. We also confirm that *Crepicephalus* does indeed become extinct before the biomere boundary at Beartooth Butte and Cathedral Cliffs, i.e. its early disappearance is not due to the Signor–Lipps effect. Our results also indicate that *Cheilocephalus* almost certainly appeared (migrated in) later than *Glaphyraspis* in both sites.

We often wish to exclude taxa that are rare. Our feeling is that we can exclude taxa more or less at whim, and still keep a valid model. However, we cannot combine taxa unless they have the same appearance and extinction times. Combining taxa does happen naturally when specimens are identified at higher taxonomic levels than species. The prior distribution of the a_{jk} should change if we exclude taxa, as we typically exclude taxa with low values of a_{jk} compared with taxa that remain in the sample, but our priors are not specified with such attention to detail that this issue has affected us yet.

Convergence of our computational algorithms is an issue with our model with the current data. It is caused by a lack of identifiability of some functions of the parameters. For Beartooth Butte and Cathedral Cliffs, one set of taxa *Aphelaspis*, *Glaphyraspis* and *Cheilocephalus* coexist with each other, and when they are extant, none of *Tricrepicephalus*, *Coosina* or *Crepicephalus* exist. Conversely, when *Tricrepicephalus*, *Coosina* and *Crepicephalus* exist, none of the first set of three exist. The ratio of abundances $a_{Aph.}/a_{Gla.}$ is well defined because the two taxa coexist and we can estimate the abundance ratio based on the data. In contrast, when two taxa do not coexist and they do not coexist at different times with some third taxon, the ratio of abundances is not estimable from the data, and so, for example, $a_{Aph.}/a_{Tri.}$ is not well estimated. This means that the ratio $a_{Aph.}/a_{Tri.}$ can vary substantially in our inferences. This does not affect the fit of the model to the data, because we never predict that *Aphelaspis* and *Tricrepicephalus* coexist, and what we estimate for the relative population sizes is irrelevant and this variation has no implications for any observations we might take or predictions we might make.

Table 10.8 gives two disparate abundance values a_1 and a_2 that are well supported by the data. Suppose that we were to sample a single specimen while only *Aphelaspis*, *Glaphyraspis* and *Cheilocephalus* were extant. The last two columns in the top part of Table 10.8 give the probabilities that we sample *Aphelaspis*, *Glaphyraspis* and *Cheilocephalus* respectively for a_1 and for a_2. We see that the probabilities do not differ substantively. The bottom half of the table gives the same probabilities for *Tricrepicephalus*, *Coosina* and *Crepicephalus* and again the relative probabilities among these three taxa are very well determined by the data. We conclude that the lack of identifiability of the posterior and consequent lack of convergence has not hurt us with respect to our inferences about observable events.

Acknowledgements

This work was partially supported by NSF grant EAR-0000385 (CRM, REW).

Appendix A: Computation

The posterior distribution of this model is analytically intractable. We use Markov chain Monte Carlo (MCMC) sampling (Gilks et al. 1996) to simulate draws from the posterior distribution.

Table 10.8. Predictions for Beartooth Butte from two different abundance estimates. Columns a_1 and a_2 give two very different values of the abundances that are
supported by the model. Suppose that we were to sample a single specimen. In the
top half of the table, the column labelled π_1 gives relative probabilities of observing
Aphelaspis (*Aph.*), *Glaphyraspis* (*Gla.*) and *Cheilocephalus* (*Che.*) if they alone were
extant, and assuming column a_1 are the abundances. Similarly the column headed
π_2 is based on the abundances a_2. In the bottom half of the table we have the same
probabilities, but assuming that only *Tricrepicephalus* (*Tri.*), *Coosina* (*Con.*) and
Crepicephalus (*Cre.*) are extant. The probabilities of observing the taxa differ only
by at most 2 in the third decimal place.

Taxon	a_1	a_2	π_1	π_2
Aph.	0.006769	0.793596	0.801	0.802
Gla.	0.001197	0.139171	0.142	0.141
Che.	0.000486	0.056984	0.058	0.058
Tri.	0.331494	0.003425	0.334	0.334
Con.	0.008063	0.000097	0.008	0.009
Cre.	0.651990	0.006727	0.658	0.656

The parameters are a_k, $\alpha_k = (\alpha_{1k}, \ldots, \alpha_{Ik})^t$, and $\omega_k = (\omega_{1k}, \ldots, \omega_{Ik})^t$,
$k = 1, \ldots, K$ and also A_{00} for the exchangeable model. For all three priors, we
sample various blocks of parameters within a site, before going to the next site.
For the exchangeable prior, we then sample A_{00} and for the equal a model, we
last sample a. For the a_ks, a and A_{00}, we convert to the logit scale, and update
each coordinate with a Metropolis–Hastings (MH) step using $N_6(0, \sigma^2 I)$, with
$\sigma = 0.1$ and renormalizing a_k before accepting/rejecting.

For α_k and ω_k, we randomly choose between a single-block and a multiple-
step updating scheme. In the first scheme, used with probability 0.8, we update
(α_k, ω_k) jointly, using candidates chosen from a distribution which does not
depend on the current values of α_k or ω_k. The proposal distribution is proportional to a geometric over bins with $p = 0.5$. For example, for an appearance
we put probability mass proportional to $1/2$ on the bin closest to the first
time taxon i was observed at site k, then $(1/2)^2$ for the next closest bin and
so on until we run out of bins. The probabilities are normalized to sum to 1.
We set up the candidate distribution for all α_{ik} and ω_{ik} in a similar fashion,
extinctions starting at the bin after the last appearance. We then update all
$(\alpha_{ik}, \omega_{ik})$ or none in a single Metropolis accept/reject step. Our second updating scheme is selected with probability 0.20. We update pairs $(\alpha_{ik}, \omega_{ik})$
using the same candidate distribution, cycling through all taxa $i = 1, \ldots, I$.
This second scheme allows for more jumps than the first, but may not fully
mix the posterior, while the first can definitely reach all parts of the posterior.

We end up with MCMC samples of size 200,000 after a burn-in of 20,000.
Every fifth draw is recorded and results reported are based on these 40,000
recorded draws. This takes us roughly 7 hours in R on a 1.5 GHz Pentium IV
(for information about R see http://www.r-project.org).

Appendix B: Technical Comment on the Prior

The priors for D_{ik} and B_{ik} are both exponential from baselines t_{1k} and $t_{J_k k}$, with $p(D_{ik})$ looking forward in time, while $p(B_{ik})$ looks backwards in time. This prior gives weight to the possibility $t_{1k} \leq D_{ik} < B_{ik} \leq t_{J_k k}$, but since this event has zero likelihood, it does not affect the posterior. This prior does require that we analyse taxa that are actually extant in the data sets we study. If we wish, we can adjust the prior by putting in the constraint $B_{ik} < D_{ik}$ followed by renormalization. This adjustment does not affect the posterior, but if, for example, we wish to calculate the marginal density of the data as would be needed to make Bayes factor comparisons of two models, we will need to think carefully about the prior specification.

References

Alvarez, L. W. (1984). Experimental evidence that an asteroid impact led to the extinction of many species 65 million years ago. *Proceedings of the National Academy of Sciences*, **80**, 627–642.

Gilks, W., Richardson, S. and Spiegelhalter, D. (eds.) (1996). *Markov chain Monte Carlo in practice*. Chapman and Hall, London.

Jin, Y. G., Wang, Y., Wang, W., Shang, Q. H., Cao, C. Q. and Erwin, D. H. (2000). Pattern of marine mass extinction near the Permian–Triassic boundary in South China. *Science*, **289**, 432–436.

Marshall, C. R. (1994). Confidence intervals on stratigraphic ranges: partial relaxation of the assumption of a random distribution of fossil horizons. *Paleobiology*, **20**, 459–469.

Marshall, C. R. (1995). Distinguishing between sudden and gradual extinctions in the fossil record: Predicting the position of the Cretaceous–Tertiary iridium anomaly using the ammonite fossil record on Seymour Island, Antarctica. *Geology*, **23**, 731–734.

Marshall, C. R. (1997). Confidence intervals on stratigraphic ranges with non-random distributions of fossil horizons. *Paleobiology*, **23**, 165–173.

Marshall, C. R. (1998). Determining stratigraphic ranges. In S. K. Donovan and C. R. C. Paul (eds.), *The adequacy of the fossil record*, Wiley, London, 23–53.

Marshall, C. R. and Ward, P. D. (1996). Sudden and gradual molluscan extinctions in the latest Cretaceous in western European Tethys. *Science*, **274**, 1360–1363.

Palmer, A. R. (1965). Biomere – a new kind of biostratigraphic unit. *Journal of Paleontology*, **39**, 149–153.

Palmer, A. R. (1984). The biomere problem; evolution of an idea. *Journal of Paleontology*, **58**, 599–611.

Palmer, A. R. (1998). A proposed nomenclature for stages and series for the Cambrian of Laurentia. *Canadian Journal of Earth Sciences*, **35**, 323–328.

Paul, C. R. C. (1982). The adequacy of the fossil record reconsidered. In K. A. Joysey and A. E. Friday (eds.), *Problems of phylogenetic reconstruction*, Academic Press, London, Systematics Association Special Volume 21, 75–117.

Raup, D. M. (1989). The case for extraterrestrial causes of extinction. *Philosophical Transactions of the Royal Society of London Series B*, **325**, 421–435.

Sepkoski, J. J., Jr. (1997). Biodiversity: past, present, and future. *Journal of Paleontology*, **71**, 533–539.

Shaw, A. B. (1964). *Time in stratigraphy*. McGraw-Hill, New York.

Signor, P. W. and Lipps, J. H. (1982). Sampling bias, gradual extinction patterns, and catastrophes in the fossil record. In L. T. Silver and P. H. Schultz (eds.), *Geological implications of large asteroids and comets on the Earth*, Geological Society of America Special Paper, vol. 190, 291–296.

Solow, A. R. (1996a). Bayesian methods for inference about extinction times. In *Proceedings of the section on Bayesian statistical sciences*, American Statistical Association, Alexandria, VA, 105–108.

Solow, A. R. (1996b). Tests and confidence intervals for a common upper endpoint in fossil taxa. *Paleobiology*, **22**, 406–410.

Solow, A. R. and Smith, W. (1997). On fossil preservation and the stratigraphic ranges of taxa. *Paleobiology*, **23**, 271–277.

Solow, A. R. and Smith, W. (2000). Testing for a mass extinction without selecting taxa. *Paleobiology*, **26**, 647–650.

Springer, M. S. (1990). The effect of random range truncations on patterns of evolution in the fossil record. *Paleobiology*, **16**, 512–520.

Stitt, J. H. (1971). Repeating evolutionary pattern in Late Cambrian trilobite biomeres. *Journal of Paleontology*, **45**, 178–181.

Strauss, D. and Sadler, P. M. (1989). Classical confidence intervals and Bayesian probability estimates for ends of local taxon ranges. *Mathematical Geology*, **21**, 411–427.

Thomas, R. C. (1993). *The Marjumiid–Pterocephaliid (Upper Cambrian) mass extinction event in the Western United States*. Ph.D. thesis, University of Washington, Washington, USA.

Thomas, R. C. (1995). Cambrian mass extinction 'Biomere' boundaries: a summary of thirty years of research. *Northwest Geology*, **24**, 67–75.

Vaziri, M. R. (1997). *Patterns of microfaunal occurrence across the Cenomanian–Turonian boundary in England*. Ph.D. thesis, University of Liverpool, Liverpool, UK.

Wang, S. C. and Marshall, C. R. (in preparation). Improved confidence intervals for estimating the position of mass extinction boundaries.

Weiss, R. E. and Marshall, C. R. (1999). The uncertainty in the true end point of a fossil's stratigraphic range when stratigraphic sections are sampled discretely. *Mathematical Geology*, **31**, 435–453.

11

Taking Bayes Beyond Radiocarbon: Bayesian Approaches to Some Other Chronometric Methods

Andrew R. Millard

Summary. Building on the models for, and practical applications of, Bayesian chronological analysis on the basis of radiocarbon and archaeomagnetism outlined by Buck (Chapter 1), Bayliss and Bronk Ramsey (Chapter 2) and Lanos (Chapter 3), this chapter looks forward to the potential of Bayesian modelling for other chronometric methods. Suitable mathematical formulations are suggested for dendrochronology, uranium-series, amino-acid racemization and trapped charge (luminescence and ESR) dating methods. Some are only initial suggestions for formulations which may lead to practical implementations. For dendrochronology, uranium-series and ESR dating the chapter offers more detailed models, initial implementations, and for the latter two, illustrative case studies which indicate the nature of inferences we can expect to make if such models are adopted more widely. By suggesting such a broad range of extensions to the Bayesian chronological framework this chapter offers great potential for substantially extending the kinds of problems that can be tackled within it, and provides encouragement to researchers who rely on a wide range of dating methods and would like to integrate them all within one coherent framework.

11.1 Introduction

Archaeologists and geologists have recognized for many years that chronometric dates and stratigraphic information need to be combined, if only at the level of establishing that the ordering of samples according to stratigraphy agrees with their ordering by chronometry. The last fifteen years have seen the development and routine application of Bayesian chronological modelling as means to go beyond this and actually use the stratigraphic (or other) prior chronological information to constrain and inform the quantitative estimates of time from chronometric measurements (see Chapter 1). The basic idea is a simple one and may be expressed as a form of Bayes' theorem

$$p(\text{dates}|\text{chronometry}) \propto p(\text{chronometry}|\text{dates}) \times p(\text{dates})$$

where p(dates) expresses our prior beliefs about the dates of events before obtaining chronometric measurements, p(chronometry|dates) is the likelihood which uses a model to express the probability of obtaining our observations if the dates were known and p(dates|chronometry) expresses our posterior beliefs incorporating our prior beliefs and the data. The prior can include statements about relative ordering of events, and thus incorporate stratigraphic information. In fact a small number of simple components can be combined to represent almost any stratigraphic relationship, just as in a Harris diagram the stratigraphy of an archaeological site is recorded entirely as known earlier than/later than relationships or a lack of knowledge of temporal relation (Harris 1989, see also Chapter 6, particularly Figure 6.1). The mathematical models so constructed can also incorporate extra parameters, e.g. start or end dates for phases, and more sophisticated models of the type of processes generating the dated samples. For details of the current state of the art in terms of methodology and application see Chapters 1 and 2.

As these two chapters show, the major application of Bayesian chronological models has been to the calibration and analysis of radiocarbon dates. However examination of any textbook on dating techniques in archaeology or Quaternary science (e.g. Aitken 1990; Smart and Frances 1991; Taylor and Aitken 1997) will show that a whole host of other chronometric techniques are available – and applied – within archaeology and Quaternary science. The beauty of Bayesian chronological models is their ability to combine many different types of information in a mathematically rigorous and philosophically satisfying way. It is natural therefore to extend the method to the other available dating techniques. As discussed above, in most cases, the chronometric information is incorporated into the model via the likelihood, whilst the prior chronological knowledge is incorporated via the prior. Thus to extend Bayesian chronological modelling to other techniques we need to develop suitable formulations for their likelihoods.

Some work has already been done in applying Bayesian methods to dating techniques other than radiocarbon. Buck et al. (1996, Chapter 12) have developed a Bayesian approach for the production of dendrochronological dates, given a master chronology. Millard (2002) has described a method to estimate dendrochronological felling dates when the date of some rings is known, but some or all of the sapwood is missing from the timber. Zink (2002) has used a Bayesian approach to aspects of luminescence authenticity testing. Both *OxCal* (Bronk Ramsey 1995, 1998) and *BCal* (Buck et al. 1999) provide features which allow the incorporation of any date into a model by expressing it in calendar years, but this requires an assumption that the chronometric determinations are independent of one another (as is frequently assumed in calculations involving radiocarbon) and for many techniques this is not true. *OxCal* offers a facility for incorporating luminescence dates which allows for a single correlated error term in a group of dates, but, as discussed in more detail below, a hierarchy of error terms needs to be considered. None of these applications of a Bayesian paradigm beyond radiocarbon dating have seen widespread or routine application. The best developed application is probably in archaeomagnetic dating, in which the mathematics has parallels with both radiocarbon calibration and the hierarchy of error terms for trapped charge techniques (Lanos 2001, and this volume Chapter 3; see also Section 11.4 below).

In this chapter I try to lay out some basic groundwork for the extension of Bayesian chronological analyses to dendrochronology, uranium-series dating, luminescence and ESR dating and amino-acid racemization dating. Although this list covers the major Quaternary chronometric techniques (with the exception of potassium–argon dating), the range and depth of consideration are variable. This reflects my own research interests, and the extent to which I have considered the methods, as my belief in the need for this type of work has grown over the last three years or more.

11.2 Dendrochronology

Dendrochronology, or tree-ring dating, relies on the variation in width of annual tree rings. The variation from year to year is climatically driven, and a period of 50–100 years will show a distinctive pattern. By using recent wood of known felling date, and matching its ring-width patterns with successively older pieces of wood a master chronology is built up showing how ring-widths have varied through time. The longest of these master chronologies now extend back approximately 11 ka (Spurk et al. 1998), but as each chronology is species and region specific the time range covered varies considerably. Wood of unknown age, e.g. from an archaeological site or a standing building, can have its ring-width sequence matched to a master chronology and thus a date obtained for it. The date of interest is usually that of the felling of the tree, so if the outer rings are missing an estimate must be made of the number of missing rings.

11.2.1 Matching a Timber to a Master Chronology

In general the procedures for forming master chronologies and for matching ring sequences are conducted using tree-ring indices (Cook 1990). A tree-ring index is a standardized tree-ring width which is used to remove features such as growth trend (rings often get narrower as a tree gets older). The series of tree-ring indices is statistically modelled as a time-series of constant mean and variance, with each tree having a set of variance components, one of which is the common climatic signal of interest, with others due to various causes (Cook 1990). Litton and Zainodin (1991) consider this model and derive the following results. The basic model is

$$x_{ij} = \phi_i + e_{ij}$$

where x_{ij} is the index for the jth tree in year i, $\phi_i \sim N(0, \zeta^2)$ is the climatic signal and $e_{ij} \sim N(0, \tau^2)$ represents the other components of variability and is referred to as noise. The ratio $R = \zeta^2/\tau^2$ is the signal-to-noise ratio. The expected correlation coefficient between two tree-ring index sequences when correctly aligned can then be shown to be $\rho = \sqrt{R/(R+1)}$, whilst for incorrectly aligned sequences it is zero. If the observed correlation coefficient is r, then it is convenient to work with Fisher's z-transformation, as it can be approximated by a Normal distribution. z is given by

$$z = \tfrac{1}{2} \ln \left(\frac{1+r}{1-r} \right).$$

For correct alignments z has mean

$$\nu = \tfrac{1}{2} \ln \left(\frac{1+\rho}{1-\rho} \right)$$

and for incorrect alignments it has mean zero. In both cases the variance is $1/(n-3)$ where n is the number of aligned indices.

For the Bayesian dating of an index sequence (from a single timber or a site master) Buck et al. (1996, 342ff) develop this into a Bayesian approach by forming a likelihood (here rearranged slightly and in a simplified form)

$$z|\theta \sim N\left(\nu, \sqrt{1/(n-3)}\right)$$

and a prior $\theta \sim U(t_{min}, t_{max})$, where t_{min} to t_{max} is the range of possible alignment dates. This is a vague prior where all possible dates are equally likely, but it does not include the possibility that the sequence does not match the master anywhere, because its origin is spatially or temporally outside of the master chronology, or it has experienced some unusual growing conditions. This is problematic, particularly from the viewpoint of practising dendrochronologists who reject many timbers as not datable. It seems likely that this feature of the method of Buck et al. (1996) has contributed to its lack of take-up.

However there is a way to incorporate the possibility of no match. A revised likelihood is readily derived from the above and the work of Litton and Zainodin (1991):

$$z|\theta, M \sim N\left(M\nu, \sqrt{1/n-3}\right)$$

where $M = 1$ if there is a match and $M = 0$ if there is no match to the master chronology. The prior

$$M \sim \text{Bernoulli}\,(c\,(\theta)) \qquad \text{and} \qquad \theta \sim U\,(t_{min}, t_{max}),$$

where $c(\theta)$ is a prior estimate of the probability that a timber can be dated and may depend on θ (for example due to changing amounts of imported timber through time), would allow us to form posterior estimates of the date θ and the probability that there is no match. Traditionally timbers from one site are dated by matching them to one another and then comparing a site master (consisting of the mean ring indices for the site) with a regional master chronology. However, this leads to a loss of information about the *distribution* of z-scores obtained when the timbers of the site are compared with the regional master. When accounting for the possibility of no match, the above formulation suggests that we may be better off comparing each individually with the regional master with appropriate conditional independence assumptions and allowing the distribution of z-scores to determine the posterior estimate of the probability that there is no match. The choice of the value of c is clearly important and needs some work eliciting information from practising dendrochronologists.

11.2.2 Forming a Master Chronology

The above matching process assumes a suitable master chronology is available and that in the right circumstances a site master can be formed. Currently the most rigorous method available to form masters is the 'objective method' of Laxton et al. (1988). This provides an algorithm for matching sequences and selecting the best match within a group of timbers to form site masters which can then be combined in the same way to form a regional master. The method is based upon finding the best alignments of pairs of timbers and seeking groups which align in such a way that the number of pairwise best alignments is maximized, with priority given to the better pairwise matches. The method is objective in that it uses only calculations from the tree-ring measurements and stated rules to form a master. From a Bayesian perspective this fails to use information, e.g. an estimate of the age of a building, which could be used as prior information in constructing a master. In addition the resulting masters consist of mean ring indices, and do not carry any information on the uncertainty of these means, which might be incorporated into the matching approach described above.

Although a Bayesian, fully probabilistic approach would be desirable, there are likely to be computational problems in finding the few *a posteriori* likely alignments, as they are widely separated in the parameter space. One possible solution would be to create a hierarchy of posterior distributions and use simulated tempering MCMC, alternatively the use of prior information about the ordering of samples might help to ease the computation (G. Nicholls personal communication).

11.2.3 Sapwood Estimates

Trees are composed of inner heartwood rings and an outer set of sapwood rings. When the sapwood on a timber is missing or partly missing, the number of missing rings must be estimated to obtain a felling date. It is known that the distribution of numbers of sapwood rings in a population of trees can be approximated by a discretized log-normal distribution (Hughes et al. 1981; Hillam et al. 1987), and the parameters of the distribution for a region can be estimated from examination of timbers with complete sapwood (Miles

1997). This distribution can be used as a prior estimate of the number of sap-wood rings originally in a tree, before we observe the timber derived from the tree, and a posterior estimate is formed from the number of rings observed in the incomplete sapwood (Millard 2002). However the use of a discretized log-normal distribution, just because it fits the data, seems unsatisfactory. It would be better to have a distribution derived from a physiological considera-tion of how much sapwood a tree needs, although in terms of changes to 95% posterior confidence intervals, the gain in using such a model is likely to be very small.

11.3 Uranium-series Dating

11.3.1 Closed Systems

Uranium-series dating is based upon the geochemical separation of uranium (U) and its decay products (progeny), deposition of that U in a new mineral and subsequent build up of the progeny isotopes. The primary decay routes of interest in archaeological and Quaternary dating applications are

$$
\underset{t_{\frac{1}{2}}=4.5\text{Ga}}{^{238}\text{U}} \xrightarrow{\alpha} \underset{t_{\frac{1}{2}}=24\text{days}}{^{234}\text{Th}} \xrightarrow{\beta} \underset{t_{\frac{1}{2}}=1\,\text{min}}{^{234}\text{Pa}} \xrightarrow{\beta} \underset{t_{\frac{1}{2}}=248\text{ka}}{^{234}\text{U}} \xrightarrow{\alpha} \underset{t_{\frac{1}{2}}=75.2\text{ka}}{^{230}\text{Th}} \xrightarrow{\alpha} \ldots
$$

and

$$
\underset{t_{\frac{1}{2}}=713\text{Ma}}{^{235}\text{U}} \xrightarrow{\alpha} \underset{t_{\frac{1}{2}}=34.3\text{ka}}{^{231}\text{Pa}} \xrightarrow{\alpha} \ldots
$$

Dates based on these two different isotopic systems are known as uranium–thorium (U-Th) and uranium–protactinium (U-Pa) dates. Protactinium (Pa) and thorium (Th) are insoluble in water under most natural circumstances, whilst U is soluble in oxidizing conditions, so the separation is often and easily obtained. The essential assumptions of the method are

- no Th or Pa present at the time of formation of the mineral,
- no gain or loss of U, Th or Pa after formation, except by radioactive decay.

A system satisfying these conditions is called a closed system. Equations for the activity ratios of various isotopes at time t from formation are readily derived (Ivanovich and Harmon 1992, Appendix A)

$$
\frac{^{230}\text{Th}}{^{234}\text{U}} = \left(\frac{^{238}\text{U}}{^{234}\text{U}}\right)(1 - \exp(-\lambda_0 t))
$$

$$
+ \left(1 - \frac{^{238}\text{U}}{^{234}\text{U}}\right)\frac{\lambda_0}{\lambda_0 + \lambda_4}(1 - \exp(-(\lambda_0 - \lambda_4)t))
$$

$$
\frac{^{231}\text{Pa}}{^{235}\text{U}} = \exp(-\lambda_1 t)
$$

where each isotope symbol represents the activity of that isotope (i.e. the rate of radioactive disintegration per unit mass of sample) and λ_n is the decay constant for the isotope of mass $230 + n$. Typically corals and detritus-free cave deposits can be dated with these equations. The isotope ratios are determined with uncertainties. Here I follow the common implicit assumption in the literature that they can be approximated as normally distributed errors (Ludwig 2001, p. 21, is a rare explicit statement of the approximation). Hence likelihoods suitable for incorporation in chronological models for U-Th dating are

$$r_{04}|\theta \sim N\left(\mu_{04}, s_{04}\right)$$
$$r_{84} \sim N\left(\mu_{84}, s_{84}\right)$$
$$\mu_{04} = \mu_{84}\left(1 - \exp\left(-\lambda_0\theta\right)\right)$$
$$+ \left(1 - \mu_{84}\right)\frac{\lambda_0}{\lambda_0 + \lambda_4}\left(1 - \exp\left(-\left(\lambda_0 - \lambda_4\right)\theta\right)\right) \quad (11.1)$$

and for U-Pa dating

$$r_{15} \sim N\left(\mu_{15}, s_{15}\right)$$
$$\mu_{15} = \exp\left(-\lambda_1\theta\right)$$

where r_{nm} is the observed ratio of isotope $230 + n$ to $230 + m$, μ_{nm} is the true underlying value and s_{nm} is the observed standard deviation. A prior on θ is specified as usual, and a suitable prior is placed on μ_{84}, e.g. $\mu_{84} \sim U(0, 20)$, would be a vague prior encompassing the entire range of values observed in natural systems. Problems arise however in that for a given μ_{84} there is an upper limit on μ_{04} corresponding to an infinitely old sample, and when r_{04} is measured within error of this value and combined with an improper vague prior $\theta \sim U(0, \infty)$, the posterior is improper (in the literature using likelihood-based calculations this is often expressed as a 'greater than' age). Changing to a vague proper prior such as $\theta \sim U(0, M)$ results in posterior ranges strongly dependent on M. With modern, precise, TIMS (thermal ionization mass spectrometry) measurements this is less of a problem than with older, less precise, α-spectrometry measurements. However the high precision of some TIMS measurements leads to a further complication, in that the uncertainty in the decay constants is no longer negligible and must be accounted for. Such uncertainties are of course common to all U-series dates in a chronological model, and the independence assumption, which simplifies the calculations in radiocarbon chronological models, cannot be maintained.

11.3.2 Open Systems

There are many attractive sample materials with high uranium contents and clear archaeological relevance that do not conform to closed system assumptions. The assumptions break down in two situations:

- where some or all of the U is taken up post-depositionally (e.g. bones, teeth and shells),

- where the initial Th or Pa content is not negligible (e.g. detritus-rich calcite deposits).

Before these can be included in any chronological models (Bayesian or otherwise) the processes of uptake or detrital contamination need to be understood.

Various equations have been proposed for uranium uptake in shells and bone, mostly by finding convenient mathematical forms for the calculations. (For a brief overview and more detailed references see Millard and Pike 1999.) Where such a suitable mathematical form is chosen (e.g. U concentration increasing at a constant rate, which is known as linear uptake), analytical forms for the dependence of isotope ratios on time can be derived and used in a similar way to the forms above. For all of the published models there are some apparently successful dating applications and others which have failed, implying that in the latter cases the convenient mathematical forms are not representative of physical reality. One model is based on a physico-chemical description of U uptake in bone and teeth as a diffusion–adsorption process (Millard and Hedges 1996); this allows rejection of samples not conforming to its assumptions and is applicable in one-third to one-half of cases (Pike and Hedges 2002). However the calculations involve the numerical solution of a differential equation and would require more work than simpler equations for MCMC Bayesian analysis.

11.3.3 Detritally Contaminated Samples

Detrital contamination is often tackled using an isochron approach. Several sub-samples are taken, either mechanically or by sequential leaching with acid, and measurements made on them. Given that they are of the same age, but usually contain varying amounts of detritus, they yield differing isotope ratios which can then be used in an appropriate regression to determine the composition when there is zero detritus. Ludwig and Titterington (1994) discuss the various ways in which this can be approached, and maximum-likelihood estimation of the parameters of interest, together with error estimates. One representation of the regression is

$$\mu_{04i} = \mu_{04c} + B\mu_{24i} \tag{11.2}$$
$$\mu_{48i} = \mu_{48c} + C\mu_{24i} \tag{11.3}$$

where i subscripts refer to sub-samples and c subscripts to ratios in the mineral common to all sub-samples whose formation is to be dated. B and C are combinations of isotope ratios which are constant for any one sample, but, if required, may be used to estimate the isotope composition of the detritus. μ_{04c} is given in terms of μ_{48c} and θ by Equation 11.1. Prior probabilities must be specified for B, C, μ_{48c} and μ_{24i}.

This procedure may be illustrated by re-analysis of data from a laminated stalactite which grew in Soreq Cave, Israel (Kaufman et al. 1998).

Stalactites and stalagmites from this cave give a long record of palaeoclimatic changes in the Near East from their stable isotope composition and have been extensively uranium-series dated by both alpha-spectrometry and mass-spectrometry. Here I reanalyse the mass-spectrometric uranium-series data from stalactite 2-N. I take a prior of $\mu_{48c} \sim U(0, 20)$ and vague priors of $N(0, 1000)$ on B, C, and μ_{24i} for each lamina. For those lamina with more than one uranium-series measurement the isochron Equation 11.2 is used. Following the original authors, for those laminae with only one measurement a detrital isotope composition of $^{230}\text{Th}/^{234}\text{U} = ^{234}\text{U}/^{238}\text{U} = 1$ and $^{232}\text{Th}/^{234}\text{U} = 0.542 \pm 0.030$ is assumed, which allows the isochron to be created with this as the detrital endpoint. Table 11.1 shows the results of calculating dates using the above methodology, first assuming that there is no information about the ordering of the samples, and a vague prior for each lamina of $\theta \sim U(0, 10^7)$ and, secondly, introducing the additional constraint that the dates of the laminae must reflect the order of their growth, so that $\theta_A < \theta_B < \theta_C < \cdots < \theta_K$. Growth rates for each segment are estimated by taking the difference in depth and dividing by the difference in age. The first model gives dates very close to those in the original paper. With an ordered prior most of the marginal 95% highest posterior density regions for the dates change only slightly (except for lamina H) but the growth rate estimates are now confined to positive values. The rates estimated with an ordered prior are more precise and give a more realistic estimate of the rates of deposition of the layers in this stalactite and thus of the resolution of the palaeoclimatic record it contains.

11.4 Luminescence and ESR Dating

Luminescence dating and ESR (electron spin resonance) dating share a common physical mechanism in the accumulation of trapped charge with time due to exposure to natural radiation, and for the purposes of the construction of chronological models differ only in the methods of measurement and in some features of the sample materials. ESR and luminescence dating depend on the determination of the natural radiation dose to which a sample has been exposed during burial (D_E), and the rate at which that dose was acquired ($\dot{D}$), so that

$$\text{age} = \frac{D_E}{\dot{D}}$$

The dose rate is the sum of a series of sources of radiation:

- the sample itself, $\dot{D}_{\text{int}}$
- the gamma radiation dose from the sediment, $\dot{D}_\gamma$
- the beta radiation dose from the sediment, $\dot{D}_\beta$
- any attached dentine, $\dot{D}_{\text{DE}}$ (only in the case of ESR dating of tooth enamel).

Table 11.1. Uranium-series data and modelling results on Stalactite 2-N from Soreq Cave, Israel. Data from Kaufman et al. (1998). Isotope ratios are given with numbers in parentheses indicating uncertainty in the last digit(s). Ages and growth rate are given as 95% highest posterior density regions.

Lamina	Depth (mm)	$^{234}U/^{238}U$	$^{230}Th/^{234}U$	$^{232}Th/^{234}U$	Simple isochron Age (ka)	Growth rate (mm/ka)	Ordered isochron Age (ka)	Growth rate (mm/ka)
K	123.5	1.0717(049)	0.1816(16)	0.0078(1)	19.77–20.68	2.7–4.7	19.80–20.70	2.8–4.8
Ja	109	1.0634(052)	0.1644(20)	0.0107(1)	15.08–17.15	−234–256	15.38–17.27	4.1–169
Jb		1.0707(045)	0.1948(21)	0.0230(1)				
Ia	93	1.0570(080)	0.1440(30)	0.0095(1)	14.16–16.99	0.9–3.6	14.18–16.50	1.3–4.0
Ib		1.0655(090)	0.1619(59)	0.0261(3)				
Ha	82	1.0358(046)	0.1239(20)	0.0242(1)	5.54–12.87	−181–172	8.01–12.57	3.2–128
Hb		1.0354(111)	0.1416(55)	0.0341(4)				
Fa	59	1.0302(097)	0.0998(17)	0.0158(1)	7.65–8.74	−40–71	7.70–8.74	3.9–46
Ea	49.5	1.0231(062)	0.0824(15)	0.0148(1)	6.54–8.09	2.7–9.0	6.59–8.04	3.0–9.2
Eb		1.0250(060)	0.1066(31)	0.0355(2)				
D	40	1.0161(063)	0.0788(25)	0.0176(9)	4.51–6.10	−126–173	4.69–6.08	5.9–98
C	28.5	1.0141(145)	0.0567(14)	0.0080(3)	4.34–5.10	6.2–11.9	4.33–5.09	6.3–12
B	17	1.0174(045)	0.0481(10)	0.0097(3)	3.05–3.72	3.9–5.1	3.07–3.73	3.9–5.1
A	6	1.0271(044)	0.0292(06)	0.0113(3)	0.60–1.23		0.62–1.23	

All of these are determined with an associated error term. $\dot{D}_{\text{int}}$ and $\dot{D}_{\text{DE}}$ are measurements with errors unique to each sample, the same is assumed here for D_{E} although there will be some systematic error in this measurement. $\dot{D}_\gamma$ and $\dot{D}_\beta$ measurements apply to groups of dates, so their errors are not independent between samples in a group. (In the literature these errors in common are often described as 'systematic', e.g. Aitken 1985, Appendix B, but this is not a particularly helpful description.) Such dependence needs to be taken into account in any analysis of dates. In the case of ESR dating of tooth enamel the values of $\dot{D}_{\text{int}}$ and $\dot{D}_{\text{DE}}$ may be uncertain due to uncertainty in the mode of uranium-uptake, but this will be sample specific. The forms used are usually early uptake (corresponding to the closed system in uranium-series dating) or linear uptake (see Section 11.3.2). More complex analyses combine ESR measurements with uranium-series measurements, to constrain the possibilities for U uptake. A more detailed treatment of luminescence and ESR dating can be found in Grün (2001).

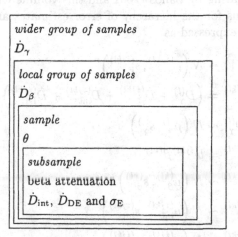

Fig. 11.1. The hierarchy of parameters in common between different dating samples. Each inner box is repeated within the box surrounding it, with different values of the parameters for different sets of samples.

A variety of measurements contribute to the evaluation of each of the components of the dose rate, the measurements differing with the sample material. However ESR dates on tooth enamel are slightly more complicated to calculate, with $\dot{D}_{\text{DE}}$ and uptake to consider in addition to the parameters of luminescence dating. Here I will consider ESR dating of tooth enamel (following the method of Millard in preparation) and the simpler case of luminescence dating is omitted. Ultimately, to obtain a general chronological modelling tool, likelihoods for all possible samples need to be considered, and in some cases this would be especially worthwhile as some of the parameters, e.g. water content of pottery, have non-normal distributions of uncertainty, although standard error-estimation procedures (e.g. Aitken 1985, Appendix B) treat them as normal.

11.4.1 A Likelihood for ESR Dating

Consideration of the components of the dose rate shows that where there are multiple samples they fall into a hierarchy of groups for these parameters (Figure 11.1). $\dot{D}_{int}$, $\dot{D}_{DE}$ and a beta attenuation factor are unique to a sub-sample; a date, θ is shared by sub-samples of the same tooth; the beta dose-rate from the sediment is common to a group of samples from the same sediment, but each experiences a different attenuation to give $\dot{D}_\beta$; the gamma dose-rate $\dot{D}_\gamma$ is homogeneous on a larger spatial scale, often for all samples from a stratum. This is expressed probabilistically by expressing the probability of the entire set of readings in a hierarchical model with conditional independence of the DE values. This model is mathematically similar to that derived for archaeomagnetic dating by Lanos (2001 and this volume Chapter 3), though the physical reasons for the hierarchy of error estimates are different. Thus the model may be expressed as

$$D_{\mathrm{E}}^{(ijkl)} \,|\, \theta \sim N\left(\mu_{\mathrm{E}}^{(ijkl)}, s_{\mathrm{E}}^{(ijkl)}\right)$$

$$\mu_{\mathrm{E}}^{(ijkl)} = \left(\dot{D}_\gamma^{(l)} + \dot{D}_\beta^{(ijkl)} + \dot{D}_{\mathrm{int}}^{(ijkl)} + \dot{D}_{\mathrm{DE}}^{(ijkl)}\right) \theta^{(jkl)}$$

$$\dot{D}_\gamma^{(l)} \sim N\left(\mu_\gamma^{(l)}, s_\gamma^{(l)}\right)$$

$$\dot{D}_\beta^{(ijkl)} = b^{(ijkl)} \beta^{(kl)}$$

$$\beta^{(kl)} \sim N\left(\mu_\beta^{(kl)}, s_\beta^{(kl)}\right)$$

$$\dot{D}_{\mathrm{int}}^{(ijkl)} \sim N\left(\mu_{\mathrm{int}}^{(ijkl)}, s_{\mathrm{int}}^{(ijkl)}\right)$$

$$\dot{D}_{\mathrm{DE}}^{(ijkl)} \sim N\left(\mu_{\mathrm{DE}}^{(ijkl)}, s_{\mathrm{DE}}^{(ijkl)}\right)$$

where i indexes over sub-samples of tooth j, from group k of samples with common beta dose-rate $\beta^{(kl)}$ and from group l of samples with common gamma dose-rate, μ_X is the true underlying value associated with observation $\dot{D}_X$, s_X is its measured standard deviation and $b^{(ijkl)}$ is the beta attenuation factor for each sub-sample. A prior needs to be put on $\theta^{(jkl)}$ as in any Bayesian chronological model.

11.4.2 Case Study: Border Cave

Millard (in preparation) has considered the case of the dating of Border Cave. This important site in South Africa has yielded a long sequence of Palaeolithic stone tool industries and early modern human remains, and there is great interest in the dating of these finds. The sequence is dated by various methods, including a series of 70 ESR determinations, which makes it the most detailed ESR dating sequence available (Grün and Beaumont 2001). This sequence of ESR dates is particularly amenable to analysis as the teeth in the cave have taken up hardly any uranium whilst buried, so the effect of U-uptake history on the dose rate is negligible. Following Grün and Beaumont (2001) only the early uptake dose rate estimates were used and two outlier measurements were excluded from the calculations.

The stratigraphy of Border Cave consists mostly of an alternating series of White Ashes (WA) and Brown Sands (BS), with clear boundaries, and differing modes of deposition. The stratigraphic model adopted is a simple one with no hiatus between these strata, and the samples are assumed to be equally likely *a priori* to have any date between the bounding dates of a stratum (cf. Zeidler et al. 1998). The hierarchy of beta and gamma dose estimates in common was derived from the published sediment U, Th and K contents. The calculation of posterior estimates of dates from the likelihood and prior were implemented using Markov chain Monte Carlo methods via *WinBUGS* software (Spiegelhalter et al. 2000).

Figure 11.2 shows the posterior estimates for the dates and phase boundaries after taking into account prior knowledge of the stratigraphy. In addition, other estimates may be derived from these dates, giving age estimates for the hominid remains and for the Howieson's Poort stone tool industry (Table 11.2). These estimates are contingent upon the model adopted. The sensitivity of the results to possible mis-specification of the model can be tested by alternative models. For example, inclusion of the two outliers identified by Grün and Beaumont (2001) makes little difference. Simplifying the stratigraphic scheme to the archaeological periods rather than the excavated strata alters the results slightly, and the results in this case are more sensitive to the inclusion of the outliers, with estimates for the duration of the Howieson's Poort industry reduced slightly. An analysis in *OxCal* treating the published dates as independent and normally distributed gives very similar mean values for the estimates of the phase boundaries, but reduced uncertainties, as is to be expected when the correlations of the uncertainties of the dates are ignored.

This case study demonstrates the possibilities of analysing ESR dates. The resulting age estimates are individually more precise and dates relating directly to events of interest may be inferred. In the future, it should be possible to integrate multiple dating techniques into a single analysis. At Border Cave date estimates with explicit uncertainties for the hominid specimens are obtained, which could be used in further analyses of, for example, the chronological order of hominid fossils from several sites. The Howieson's Poort industry is earlier than some have claimed, but this analysis does not support the claim of Grün and Beaumont (2001) that at Border Cave this industry lasted 20 ka rather than 10 ka, instead it demonstrates that significant uncertainty still remains as to its duration.

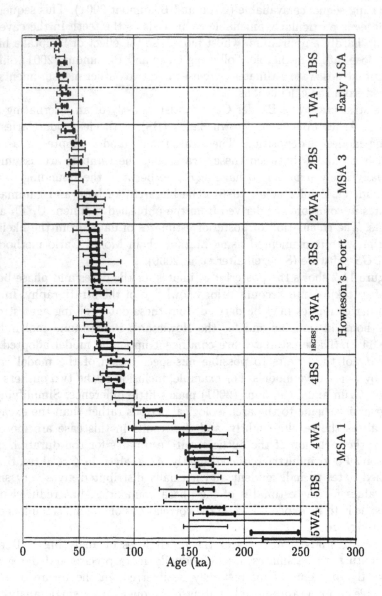

Fig. 11.2. ESR dates and modelled chronology at Border Cave. Thin black bars: dates without model from Grün and Beaumont (2001) as 95% confidence ranges; thick black bars: dates with stratigraphic model; thick grey bars: modelled phase boundary dates. Modelled dates shown as 95% highest posterior density regions. Where there are sub-samples (a, b, c) from a tooth the single modelled date for the tooth is shown under sub-sample a.

Table 11.2. Estimates of chronological parameters at Border Cave. The stratigraphic positions of the hominid remains BC 1 and 2 are not securely known.

Parameter	95% highest posterior density	Estimate by Grün & Beaumont (2001)
Start of Howieson's Poort	70–84 ka	79 ka
End of Howieson's Poort	59–66 ka	60 ka
Length of Howieson's Poort	5.8–22.2 ka	20 ka
Hominids BC 1 and 2 (if from 4BS)	74–93 ka	82±2 ka
Hominids BC 1 and 2 (if from 5BS)	159–176 ka	145–230 ka
Hominid BC3	69–93 ka	76±4 ka
Hominid BC5	64–73 ka	66±2 ka

11.5 Amino-acid Racemization Dating

Amino-acid racemization (AAR) is a chemical dating technique reliant on the transformation of the L-amino-acids found in all living organisms to their mirror image molecules, D-amino-acids. The thermodynamically stable state is an equal mixture of the two forms, but in living creatures energy is expended to maintain chemical disequilibrium. After death the biochemical maintenance systems are no longer in place and L-amino-acids are gradually transformed to D-amino-acids until equilibrium is reached. In principle the technique can be applied to any suitable proteinaceous material, but the commonest are bones, teeth and shells. Depending on the sample material and the site conditions, it may be used to date materials from a few thousand years to hundreds of thousands of years. AAR dating has been beset by problems caused by various processes which alter the D/L ratio, for example by leaching of amino-acids, and these have led to its general abandonment, unless a closed system can be demonstrated. Of the materials which maintain a closed system, ostrich egg-shell is commonly found on African Palaeolithic sites. At its simplest AAR is a process which occurs after death in a manner described by first order chemical kinetics, so that

$$\left(\frac{D}{L}\right) = \frac{r_0 - \exp\left(-(1+K)\,kt\right)}{r_0 K + \exp\left(-(1+K)\,kt\right)}$$

where $r_0 = \left[1 + \left(\frac{D}{L}\right)_{t=0}\right] / \left[1 - K\left(\frac{D}{L}\right)_{t=0}\right]$, K is a constant dependent on the amino-acid under consideration, and k is given by the Arrhenius equation $k = A\exp\left(-E_a/RT\right)$. The Arrhenius constant, A, and the activation energy E_a depend on the amino-acid, R is the gas constant, and T is the temperature. (For more details of the method see Aitken 1990 or Johnson and Miller 1997.) Forming a likelihood expression is straightforward (although assuming a normal distribution for the D/L ratio is not a valid approximation at values close to zero or one),

$$\left(\frac{D}{L}\right)\bigg|\,\theta \sim N\left(\mu_{D/L}, s_{D/L}\right)$$

$$\mu_{D/L} = \frac{r_0 - \exp\left(-(1+K)\,k\theta\right)}{r_0 K + \exp\left(-(1+K)\,k\theta\right)}$$

$$k = A\exp\left(-E_a/RT\right)$$

where $s_{D/L}$ is the measurement uncertainty. However the 'constants' r_0, A and E_a are experimentally determined and may actually have significant uncertainty that will have to be taken into account. r_0 is obtained from racemization measurements on modern materials. There are a variety of approaches to estimating k and/or T. The rate constant can be estimated by:

1. high temperature laboratory experiments to derive E_a and A;
2. use of radiocarbon (or occasionally uranium-series) dated material and an assumption of constant site temperature to estimate k;
3. a combination of data from (i) and (ii) to derive E_a and A.

However this is all subject to an assumption of constant temperature. Such an assumption might be a suitable approximation in the Holocene, but for longer timescales temperature variation must be taken into account. This may be done by:

1. an assumption of temperature which changes stepwise by an estimated amount at glacial/interglacial transitions;
2. site temperature reconstructed from a suitable palaeotemperature record and some model for the differences between the site of the temperature record and the site to be dated.

If site temperature is constant or changes stepwise and E_a and A are known then a model can be readily formulated. Where there is calibration of the rate constant from radiocarbon dates then the implementation will have to incorporate radiocarbon calibration, and the posterior AAR age estimates will have multi-modal probability distributions. More complex, but demonstrably better in its dating results, is the approach based on a palaeotemperature reconstruction (M. Collins, personal communication). To implement this requires numerical integration of the differential equations from which the constant temperature dating equation is derived. For a full analysis of the uncertainties the temperature reconstruction model would need to be considered to allow its uncertainties to be included. These uncertainties may well be the major ones in the estimation of dates, as a 1°C increase in the temperature estimate can lead to a 20% decrease in the date (Miller et al. 2000).

11.6 Conclusions

It is clear that the approaches and equations for all the dating techniques discussed above can be formulated in a Bayesian fashion, and thus they can be incorporated into chronological models. The actual calculations, particularly if they are to be conducted by MCMC, will need to be considered in detail in order to ensure that they are fast enough and reliable enough to produce useful results.

Application of these methods will require further collaboration between statisticians and chronometrists and, if published data are to be amenable to analysis by these methods, publication of full details of the chronometry. With ESR dating this is usually the case (as at Border Cave), but AAR dates are frequently published as mean and standard deviations for the population of samples from a stratum (and this prevents their incorporation into my analysis of the dating at Border Cave). With these requirements satisfied there are many applications in archaeology and Quaternary science that could benefit from Bayesian chronological modelling which formally integrates all available chronometric and relative dating information to yield chronologies which are as complete and as robust as possible.

References

Aitken, M. J. (1985). *Thermoluminescence dating*. Academic Press, London.

Aitken, M. J. (1990). *Science-based dating in archaeology*. Longman, London.

Bronk Ramsey, C. (1995). Radiocarbon calibration and analysis of stratigraphy: the OxCal program. *Radiocarbon*, **37**, 425–430.

Bronk Ramsey, C. (1998). Probability and dating. *Radiocarbon*, **40**, 461–474.

Buck, C. E., Cavanagh, W. G. and Litton, C. D. (1996). *Bayesian approach to interpreting archaeological data*. John Wiley, Chichester.

Buck, C. E., Christen, J. A. and James, G. N. (1999). BCal: an on-line Bayesian radiocarbon calibration tool. *Internet Archaeology*, **7**.
URL http://intarch.ac.uk/journal/issue7/buck/

Cook, E. (1990). A conceptual linear aggregate model for tree rings. In E. Cook and L. A. Kairiukstis (eds.), *Methods of dendrochronology: applications in the environmental sciences*, Kluwer, Dordrecht, 98–104.

Grün, R. (2001). Trapped charge dating (ESR, TL, OSL). In D. R. Brothwell and A. M. Pollard (eds.), *Handbook of archaeological sciences*, Wiley, Chichester, 47–62.

Grün, R. and Beaumont, P. (2001). Border Cave revisited: a revised ESR chronology. *Journal of Human Evolution*, **40**, 467–482.

Harris, E. C. (1989). *Principles of archaeological stratigraphy*. Academic Press, London, second edn.

Hillam, J., Morgan, R. A. and Tyers, I. (1987). Sapwood estimates and the dating of short ring sequences. In R. G. W. Ward (ed.), *Applications of tree-ring studies: current research in dendrochronology and related areas*. British Archaeological Reports, Oxford, International Series, **S333**, 165–185.

Hughes, M. K., Milsom, S. J. and Leggett, P. A. (1981). Sapwood estimates in the interpretation of tree-ring dates. *Journal of Archaeological Science*, **8**, 381–390.

Ivanovich, M. and Harmon, R. S. (eds.) (1992). *Uranium-series disequilibrium: applications to earth, environmental and marine sciences*. Oxford University Press, Oxford.

Johnson, B. J. and Miller, G. H. (1997). Archaeological application of amino acid racemization. *Archaeometry*, **39**, 265–288.

Kaufman, A., Wassserburg, G. J., Porcelli, D., Bar-Matthews, M., Ayalon, A. and Halicz, L. (1998). U-Th isotope systematics from the Soreq cave, Israel and climatic correlations. *Earth and Planetary Science Letters*, **156**, 141–155.

Lanos, P. (2001). L'approche bayésienne en chronométrie: application à l'archéomagnétisme. In J. N. Barrandon, P. Guibert and V. Michel (eds.), *Datation, XXIe rencontres internationales d'archéologie et d'histoire d'Antibes*. APDCA, Antibes, 113–139.

Laxton, R. R., Litton, C. D. and Zainodin, H. J. (1988). An objective method for forming a master ring-width sequence. In T. Hackens, A. V. Munaut and C. Till (eds.), *Wood and archaeology*. PACT, Strasbourg, Journal of the European Study Group on Physical, Chemical and Mathematical Techniques Applied to Archaeology, Conseil de l'Europe, **22**, 25–35.

Litton, C. D. and Zainodin, H. J. (1991). Statistical models of dendrochronology. *Journal of Archaeological Science*, **18**, 429–440.

Ludwig, K. R. (2001). *Users manual for Isoplot/Ex 2.49: a geochronological toolkit for Microsoft Excel*. Berkeley Geochronology Center, Berkeley.

Ludwig, K. R. and Titterington, D. M. (1994). Calculation of ^{230}Th isochrons, ages and errors. *Geochimica et Cosmochimica Acta*, **58**, 5031–5042.

Miles, D. (1997). The interpretation, presentation and use of tree-ring dates. *Vernacular Architecture*, **28**, 40–56. As amended by erratum slip in *Vernacular Architecture* **29**.

Millard, A. R. (2002). A Bayesian approach to sapwood estimates and felling dates in dendrochronology. *Archaeometry*, **44**, 137–144.

Millard, A. R. (in preparation). Bayesian analysis of ESR dates.

Millard, A. R. and Hedges, R. E. M. (1996). A diffusion–adsorption model of uranium uptake by archaeological bone. *Geochimica et Cosmochimica Acta*, **60**, 2139–2152.

Millard, A. R. and Pike, A. W. G. (1999). Uranium-series dating of the Tabun Neanderthal: a cautionary note. *Journal of Human Evolution*, **36**, 581–585.

Miller, G. H., Hart, C. P., Roark, B. and Johnson, B. J. (2000). Isoleucine epimerization in eggshells of the flightless Australian birds Genyornis and Dromaius. In G. A. Goodfriend, M. J. Collins, M. L. Fogel, S. A. Macko and J. F. Wehmiller (eds.), *Perspectives in amino acid and protein geochemistry*, Oxford University Press, New York, 161–181.

Pike, A. W. G. and Hedges, R. E. M. (2002). U-series dating of bone using the diffusion–adsorption model. *Geochimica et Cosmochimica Acta*, **66**, 4273–4286.

Smart, P. L. and Frances, P. D. (1991). *Quaternary dating methods – a users guide*. *Technical guide 4*. Quaternary Research Association, London.

Spiegelhalter, D., Thomas, A. and Best, N. (2000). *WinBUGS user manual version 1.3*. MRC Biostatics Unit, Cambridge.
URL http://www.mrc-bsu.cam.ac.uk/bugs

Spurk, M., Friedrich, M., Hofmann, J., Remmele, S., Frenzel, B., Leuschner, H. H. and Kromer, B. (1998). Revisions and extensions of the Hohenheim oak and pine chronologies: new evidence about the timing of the Younger Dryas/Preboreal Transition. *Radiocarbon*, **40**, 1107–1116.

Taylor, R. E. and Aitken, M. J. (eds.) (1997). *Chronometric and allied dating in archaeology*. Plenum, New York.

Zeidler, J. A., Buck, C. E. and Litton, C. D. (1998). The integration of archaeological phase information and radiocarbon results from the Jama River Valley, Ecuador: a Bayesian approach. *Latin American Antiquity*, **9**, 135–159.

Zink, A. J. C. (2002). Bayesian approach applied to authenticity testing by luminescence. *Archeologia e Calcolatori*, **13**, 211–216.

Index

Lecture Notes in Statistics

For information about Volumes 1 to 122, please contact Springer-Verlag

159: Marc Moore (Editor), Spatial Statistics: Methodological Aspects and Some Applications. xvi, 282 pp., 2001.

160: Tomasz Rychlik, Projecting Statistical Functionals. viii, 184 pp., 2001.

161: Maarten Jansen, Noise Reduction by Wavelet Thresholding. xxii, 224 pp., 2001.

162: Constantine Gatsonis, Bradley Carlin, Alicia Carriquiry, Andrew Gelman, Robert E. Kass Isabella Verdinelli, and Mike West (Editors), Case Studies in Bayesian Statistics, Volume V. xiv, 448 pp., 2001.

163: Erkki P. Liski, Nripes K. Mandal, Kirti R. Shah, and Bikas K. Sinha, Topics in Optimal Design. xii, 164 pp., 2002.

164: Peter Goos, The Optimal Design of Blocked and Split-Plot Experiments. xiv, 244 pp., 2002.

165: Karl Mosler, Multivariate Dispersion, Central Regions and Depth: The Lift Zonoid Approach. xii, 280 pp., 2002.

166: Hira L. Koul, Weighted Empirical Processes in Dynamic Nonlinear Models, Second Edition. xiii, 425 pp., 2002.

167: Constantine Gatsonis, Alicia Carriquiry, Andrew Gelman, David Higdon, Robert E. Kass, Donna Pauler, and Isabella Verdinelli (Editors), Case Studies in Bayesian Statistics, Volume VI. xiv, 376 pp., 2002.

168: Susanne Rässler, Statistical Matching: A Frequentist Theory, Practical Applications and Alternative Bayesian Approaches. xviii, 238 pp., 2002.

169: Yu. I. Ingster and Irina A. Suslina, Nonparametric Goodness-of-Fit Testing Under Gaussian Models. xiv, 453 pp., 2003.

170: Tadeusz Caliński and Sanpei Kageyama, Block Designs: A Randomization Approach, Volume II: Design. xii, 351 pp., 2003.

171: D.D. Denison, M.H. Hansen, C.C. Holmes, B. Mallick, B. Yu (Editors), Nonlinear Estimation and Classification. x, 474 pp., 2002.

172: Sneh Gulati, William J. Padgett, Parametric and Nonparametric Inference from Record-Breaking Data. ix, 112 pp., 2002.

173: Jesper Møller (Editor), Spatial Statistics and Computational Methods. xi, 214 pp., 2002.

174: Yasuko Chikuse, Statistics on Special Manifolds. xi, 418 pp., 2002.

175: Jürgen Gross, Linear Regression. xiv, 394 pp., 2003.

176: Zehua Chen, Zhidong Bai, Bimal K. Sinha, Ranked Set Sampling: Theory and Applications. xii, 224 pp., 2003